T0210604

Detecting Fake News on Social Media

Synthesis Lectures on Data Mining and Knowledge Discovery

Editors

Jiawei Han, *University of Illinois at Urbana–Champaign*
Johannes Gehrke, *Cornell University*
Lise Getoor, *University of California, Santa Cruz*
Robert Grossman, *University of Chicago*
Wei Wang, *University of North Carolina, Chapel Hill*

Synthesis Lectures on Data Mining and Knowledge Discovery is edited by Jiawei Han, Lise Getoor, Wei Wang, Johannes Gehrke, and Robert Grossman. The series publishes 50- to 150-page publications on topics pertaining to data mining, web mining, text mining, and knowledge discovery, including tutorials and case studies. Potential topics include: data mining algorithms, innovative data mining applications, data mining systems, mining text, web and semi-structured data, high performance and parallel/distributed data mining, data mining standards, data mining and knowledge discovery framework and process, data mining foundations, mining data streams and sensor data, mining multi-media data, mining social networks and graph data, mining spatial and temporal data, pre-processing and post-processing in data mining, robust and scalable statistical methods, security, privacy, and adversarial data mining, visual data mining, visual analytics, and data visualization.

Ensemble Methods in Data Mining: Improving Accuracy Through Combining Predictions
Giovanni Seni and John F. Elder
2010

Modeling and Data Mining in Blogosphere
Nitin Agarwal and Huan Liu
2009

Detecting Fake News on Social Media
Kai Shu and Huan Liu

ISBN: 978-3-031-00787-3   paperback
ISBN: 978-3-031-01915-9   ebook
ISBN: 978-3-031-00110-9   hardcover

DOI: 10.1007/978-3-031-01915-9

A Publication in the Springer series
SYNTHESIS LECTURES ON DATA MINING AND KNOWLEDGE DISCOVERY

Lecture #18
Series Editors: Jiawei Han, *University of Illinois at Urbana-Champaign*
              Johannes Gehrke, *Cornell University*
              Lise Getoor, *University of California, Santa Cruz*
              Robert Grossman, *University of Chicago*
              Wei Wang, *University of North Carolina, Chapel Hill*
Series ISSN
Print 2151-0067   Electronic 2151-0075

Detecting Fake News on Social Media

Kai Shu and Huan Liu
Arizona State University

SYNTHESIS LECTURES ON DATA MINING AND KNOWLEDGE DISCOVERY #18

ABSTRACT

In the past decade, social media has become increasingly popular for news consumption due to its easy access, fast dissemination, and low cost. However, social media also enables the wide propagation of "fake news," i.e., news with intentionally false information. Fake news on social media can have significant negative societal effects. Therefore, fake news detection on social media has recently become an emerging research area that is attracting tremendous attention. This book, from a data mining perspective, introduces the basic concepts and characteristics of fake news across disciplines, reviews representative fake news detection methods in a principled way, and illustrates challenging issues of fake news detection on social media. In particular, we discussed the value of news content and social context, and important extensions to handle early detection, weakly-supervised detection, and explainable detection. The concepts, algorithms, and methods described in this lecture can help harness the power of social media to build effective and intelligent fake news detection systems. This book is an accessible introduction to the study of *detecting fake news on social media*. It is an essential reading for students, researchers, and practitioners to understand, manage, and excel in this area.

This book is supported by additional materials, including lecture slides, the complete set of figures, key references, datasets, tools used in this book, and the source code of representative algorithms. The readers are encouraged to visit the book website for the latest information:

http://dmml.asu.edu/dfn/

KEYWORDS

fake news, misinformation, disinformation, social computing, social media, data mining, social cyber security, machine learning

To my parents and wife Ling.

– KS

To my parents, wife, and sons.

– HL

Contents

Acknowledgments

First and foremost, we are thankful to our colleagues in the Data Mining and Machine Learning Lab for their helpful suggestions: Deepak Mahudeswaran, Issac Jones, Justin Sampson, Tahora H. Nazer, Suhas Ranganath, Jundong Li, Liang Wu, Ghazaleh Beigi, Lu Cheng, Nur Shazwani Kamrudin, Ruocheng Guo, Kaize Ding, Raha Moraffah, Matthew Davis, Ahmadreza Mosallanezhad, and Alex Nou.

We are grateful to our collaborators for their helpful discussion on this project: H. Russell Bernard, Suhang Wang, Dongwon Lee, Juan Cao, Jiliang Tang, Reza Zafarani, Nitin Agarwal, Kathleen Carley, Sun-Ki Chai, Rebecca Goolsby, Amy Sliva, Eric Newton, and Kristy Roschke. We would like to acknowledge Morgan & Claypool, particularly the Executive Editor Diane D. Cerra for her patience, help, and encouragement throughout this project. This work is part of the projects sponsored by grants from the National Science Foundation (NSF) under grant number 1614576, the Office of Naval Research (ONR) under grant number N00014-16-1-2257.

Last but not the least, we are deeply indebted to our families for their support throughout this entire project. We dedicate this book to them, with love.

Kai Shu and Huan Liu
June 2019

CHAPTER 1

Introduction

1.1 MOTIVATION

Social media has become an important means of large-scale information sharing and communication in all occupations, including marketing, journalism, public relations, and more [179]. This change in consumption behaviors is due to some novel features such as mobility, free, and interactiveness. However, the low cost, easy access, and rapid dissemination of information of social media draw a large audience and enable the wide propagation of *fake news*, i.e., news with intentionally false information. For instance, in 2016, millions of people read and "liked" fake news stories proclaiming that Pope Francis has endorsed Donald Trump for U.S. president.[1] When the Pakistani defense minister mistakenly believed a fake news story, he threatened a nuclear war with Israel.[2] These examples clearly demonstrate that fake news stories are problematic not only for the credibility of online journalism, but also due to their detrimental real-world consequences, resulting in violence or influencing election results. Therefore, it becomes increasingly important for policy makers to regulate and discourage the creation of fake news, for online business to detect and prevent fake news, and for citizens to protect themselves from fake news.

Fake news on social media presents unique challenges. First, fake news is intentionally written to mislead readers, which makes it nontrivial to detect simply based on content. Second, social media data is large-scale, multi-modal, mostly user-generated, sometimes anonymous and noisy. Third, the consumers of social media come from different backgrounds, have disparate preferences or needs, and use social media for varied purposes. Finally, the low cost of creating social media accounts makes it easy to create malicious accounts, such as social bots, cyborg users, and trolls, all of which can become powerful sources of proliferation of fake news.

Despite the importance of the problem, our *understanding* of fake news is still limited. For example, we want to know why people create fake news, who produces and publishes it, how fake news spreads, what characteristics distinguish fake news from legitimate news, or why some people are more susceptible to fake news than others [89]. Therefore, we propose to understand fake news with disciplines such as journalism, psychology, and social science, and characterize the unique characteristics for its detection. Establishing a better understanding of fake news will allow us to come up with algorithmic solutions for *detecting* fake news and managing it before fake news is widely disseminated.

---

[1]https://www.cnbc.com/2016/12/30/read-all-about-it-the-biggest-fake-news-stories-of-2016.html

[2]https://www.nytimes.com/2016/12/24/world/asia/pakistan-israel-khawaja-asif-fake-news-nuclear.html

1.2   AN INTERDISCIPLINARY VIEW ON FAKE NEWS

In this section, we introduce the concept of fake news, the basic social and psychological theories, and a game-theoretic view of fake news, and discuss some patterns introduced by social media.

Fake News and Related Concepts   Fake news has existed for a long time, nearly around the same time as news began to circulate widely when the printing press was invented in 1439.[3] In the 1890s, Joseph Pulitzer and William Hearst competed for the same audience through sensationalism and reporting non-factual information as if they were facts, a practice that became known at the time as "yellow journalism." However, there is no agreed definition of the term "fake news." Therefore, we first discuss and compare existing widely used definitions of fake news.

A **narrow definition** of fake news is news articles that are intentionally and verifiably false and could mislead readers [6, 136, 186]. There are two key features of this definition: *authenticity* and *intent*. First, fake news includes false information that can be verified as such. Second, fake news is created with dishonest intention to mislead consumers. This definition has been widely adopted in recent studies [30, 71, 75, 95, 111]. The **broad definitions** of fake news focus on the authenticity or intent of the news content. Some papers regard satire news as fake news since the contents are false even though satire often uses a witty form to reveal its own deceptiveness to the consumers [9, 18, 59, 117]. Other literature directly treats deceptive news as fake news [116], which includes serious fabrications, hoaxes, and satires.

Definition 1.1   (NARROW DEFINITION OF FAKE NEWS) Fake news is a news article that is intentionally and verifiably false.

This narrow definition is widely adopted due to the following reasons. First, the underlying intent of fake news provides both theoretical and practical value that enables a deeper understanding and analysis of fake news. Second, any techniques for truth verification that apply to the narrow definition of fake news can also be applied to under the broader definition. Third, this definition is able to eliminate the ambiguity between fake news and related concepts that are not considered in this book.

The following concepts are different from fake news according to our definition: (1) satire news with proper context, which has no intent to mislead or deceive consumers and is unlikely to be misperceived as factual; (2) rumors that did not originate from news events; (3) conspiracy theories, which are difficult to verify as true or false; (4) misinformation that is created unintentionally; and (5) hoaxes that are only motivated by fun or to scam targeted individuals.

Psychology and Social Science   Humans are naturally not very good at differentiating between real and fake news. There are several psychological and cognitive theories that can explain this phenomenon and the influential power of fake news. Fake news mainly targets consumers by

---

[3]http://www.politico.com/magazine/story/2016/12/fake-news-history-long-violent-214535

exploiting their *individual gullibility*. The major factors which make consumers naturally gullible to fake news are as follows: (1) *Naïve Realism* [166]—consumers tend to believe that their perceptions of reality are the only accurate views, while others who disagree are regarded as uninformed, irrational, or biased; and (2) *Confirmation Bias* [99]—consumers prefer to receive information that confirms their existing views. Since these cognitive biases are inherent in human cognition, fake news can often be perceived as real by consumers.

Fake news is intended to convey inaccurate information. *Framing*, which attempts to persuade the reader by selecting some aspects of a perceived reality and make them more salient in text [38], has indivisible correlations with fake news [157]. The way people understand certain concepts depends on the way those concepts are framed. Tversky et al. [153] found that by presenting two choices in different ways, readers would largely prefer one choice over the other depending on the way they framed the choices, even though the choices were exactly the same.

Moreover, once the misperception is formed, it is very hard to have it changed. Research on cognitive psychology has theorized how people generally evaluate truth of statements. Schwarz et al. [123] propose that people usually evaluate truths using the following criteria, each of the which can be tested from both analytically and intuitively: (1) *Consensus*: Do others believe this? (2) *Consistency*: Is it compatible with my knowledge? (3) *Coherence*: Is it an internally coherent and plausible story? (4) *Credibility*: Does it come from a credible source? (5) *Support*: Is there a lot of supporting evidence? However, regardless of which truth criteria people use, easily process information has the advantage over information that is difficult to process, and is more likely to pass the criteria tests. This phenomenon explains why fake news correction may unintentionally cement the fake news they are trying to correct: when a correction attempt increases the ease with which the false claim can be processed, it also increases the odds that the false claim feels true when it is encountered again at a later point in time. A similar psychology study [100] shows that correction of fake news by the presentation of true, factual information is not only unhelpful to reduce misperceptions, but sometimes may even increase the misperceptions, especially among ideological groups.

Considering the news consumption ecosystem, we can also describe some of the social dynamics that contribute to the proliferation of fake news. Some representative social theories include the following. (1) Prospect theory describes decision making as a process by which people make choices based on the relative gains and losses as compared to their current state [61, 154]. This desire for maximizing the reward of a decision applies to social gains as well, for instance, continued acceptance by others in a user's immediate social network. (2) Social identity theory [148, 149] and normative influence theory [7, 63], this preference for social acceptance and affirmation is essential to a person's identity and self-esteem, making users likely to choose "socially safe" options when consuming and disseminating news information, following the norms established in the community even if the news being shared is fake news. Interested readers can refer to [185], which summarized a useful resources of psychology and social science theories for fake news detection.

A Game-theoretic View This rational theory of fake news interactions can be modeled from an economic game-theoretical perspective [44] by formulating the news generation and consumption cycle as a two-player strategy game. In the context of fake news, we assume there are two kinds of key players in the information ecosystem: *publishers* and *consumers*. The process of news publishing is modeled as a mapping from original source p to resultant news report a with an effect of distortion bias b, i.e., $p \xrightarrow{b} a$, where $b = [-1, 0, 1]$ indicates [*left, neutral, right*] biases that affect the news publishing process. Intuitively, this is capturing the degree to which a news article may be biased or distorted to produce fake news. The utility for the publisher stems from two perspectives: (i) short-term utility—the incentive to maximize profit, which is positively correlated with the number of consumers reached; or (ii) long-term utility—their reputation in terms of news authenticity. Utility of consumers consists of two parts: (i) information utility—obtaining true and unbiased information (usually extra investment cost needed); and (ii) psychology utility—receiving news that satisfies their prior opinions and social needs, e.g., confirmation bias and prospect theory. Both publisher and consumer try to maximize their overall utilities in this strategy game of the news consumption process. We can observe that fake news happens when the short-term utility dominates a publisher's overall utility and psychology utility dominates the consumer's overall utility, and an equilibrium is maintained. This explains the social dynamics that lead to an information ecosystem where fake news can thrive.

1.3 FAKE NEWS IN SOCIAL MEDIA AGE

In this section, we first introduce the unique characteristics brought by social media, then introduce the problem definition, principles, and methods from the perspectives of news content and social context, and finally introduce the advanced settings of fake news detection on social media.

1.3.1 CHARACTERISTICS OF SOCIAL MEDIA

We discuss some unique characteristics of fake news on social media. Specifically, we highlight the key features of fake news that are enabled by social media.

Malicious Accounts for Propaganda. While many users on social media are legitimate, some users may be malicious, and in some cases, some users are not even real humans. The low cost of creating social media accounts also encourages malicious user accounts, such as social bots, cyborg users, and trolls. A social bot refers to a social media account that is controlled by a computer algorithm to automatically produce content and interact with humans (or other bot users) on social media [41]. Social bots can become malicious entities designed specifically with the purpose to do harm, such as manipulating and spreading fake news on social media. Studies show that social bots distorted the 2016 U.S. presidential election online discussions on a large scale [14], and that around 19 million bot accounts tweeted in support of either Trump or Clinton in the week leading up to election day.[4] Trolls—real human users who aim to disrupt online communities and provoke consumers into an emotional response—are also playing an important role in spreading fake news on social media. For example, evidence suggests that there were 1,000 paid Russian trolls spreading fake news on Hillary Clinton.[5] Trolling behaviors are highly affected by people's emotions and the context of online discussions, which enables the easy dissemination of fake news among otherwise "normal" online communities [26]. The effect of trolling is to trigger people's inner negative emotions, such as anger and fear, resulting in doubt, distrust, and irrational behavior. Finally, cyborg users can spread fake news in a way that blends automated activities with human input. Usually cyborg accounts are registered as a camouflage and set computer programs to perform activities in social media. The easy switch of functionalities between human and bot offers cyborg users unique opportunities to spread fake news [28]. In a nutshell, these highly active and partisan malicious accounts on social media become the powerful sources of fake news proliferation.

Echo Chamber Effect. Social media provides a new paradigm of information creation and consumption for users. The information seeking and consumption process are changing from a mediated form (e.g., by journalists) to a more disintermediated way [33]. Consumers are selectively exposed to certain kinds of news because of the way news feeds appear on their homepage in social media, amplifying the psychological challenges to dispelling fake news identified above. For example, users on Facebook follow like-minded people and thus receive news that promote their favored narratives [113]. Therefore, users on social media tend to form groups containing like-minded people where they then polarize their opinions, resulting in an *echo chamber* effect. The echo chamber effect facilitates the process by which people consume and believe fake news due to the following psychological factors [104]: (1) *social credibility*, which means people are more likely to perceive a source as credible if others perceive the source is credible, especially when there is not enough information available to access the truthfulness of the source; and (2) *frequency heuristic*, which means that consumers may naturally favor information they hear frequently, even if it is fake news. Studies have shown that increased exposure to an idea is enough to generate a positive opinion of it [180, 181], and in echo chambers, users continue to share and consume the same information. As a result, this echo chamber effect creates segmented, homogeneous communities with very limited diversity. Research shows that the homogeneous communities become the primary driver of information diffusion that further strengthens polarization [32].

1.3.2 PROBLEM DEFINITION

Before we present the details of mathematical formulation of fake news detection on social media, we define the following basic notations.

---

[4]http://comprop.oii.ox.ac.uk/2016/11/18/resource-for-understanding-political-bots/
[5]http://www.huffingtonpost.com/entry/russian-trolls-fake-news_us_58dde6bae4b08194e3b8d5c4

- Let $\mathcal{A} = \{a_1, a_2, \cdots, a_N\}$ denote the corpus of N news pieces, and $\mathcal{P} = \{p_1, p_2, \cdots, p_K\}$ the set of publishers. For a specific news piece $a = \{\mathcal{S}, \mathcal{I}\}$, the text content consists of L sentences $\mathcal{S} = \{s_1, s_2, \cdots, s_L, \}$, with each sentence $s_i = \{w_{i1}, w_{i2}, \cdots, w_{iT_i}\}$ containing T_i words; and the visual content consists of V images $\mathcal{I} = \{i_1, i_2, \cdots, i_V\}$.

- *Social Context* is defined as a set of tuples $\mathcal{E} = \{e_{it}\}$ to represent the process of how news spread over time among n users $\mathcal{U} = \{u_1, u_2, ..., u_n\}$ and their corresponding posts $\mathcal{C} = \{c_1, c_2, ..., c_n\}$ on social media regarding news article a. Each engagement $e_{it} = \{u_i, c_i, t\}$ represents that a user u_i spreads news article a using c_i at time t. Note that users can spread news without any post contents such as sharing/liking, in which each engagement only contains the user and timestamps $e_{it} = \{u_i, t\}$.

Definition 1.2 (FAKE NEWS DETECTION) Given the social news engagements $\mathcal{E}$ among n users for news article a, the task of fake news detection is to predict whether the news article a is fake or not, i.e., $\mathcal{F} : \mathcal{E} \rightarrow \{0, 1\}$ such that

$$\mathcal{F}(a) = \begin{cases} 1, & \text{if } a \text{ is a piece of fake news,} \\ 0, & \text{otherwise,} \end{cases} \tag{1.1}$$

where $\mathcal{F}$ is the prediction function we want to learn.

1.3.3 WHAT NEWS CONTENT TELLS

Since fake news attempts to spread false claims in news content, the most straightforward means of detecting it is to check the truthfulness of major claims in a news article to decide the news veracity. Fake news detection on traditional news media mainly relies on exploring news content information. News content can have multiple modalities such as text, image, video. Research has explored different approaches to learn features from single or combined modalities and build machine learning models to detect fake news. We will introduce how to detect fake news by learning textual, visual, style features, and how to bring external knowledge to fact-check news claims.

1.3.4 HOW SOCIAL CONTEXT HELPS

In addition to features related directly to the content of the news articles, additional social context features can be derived from the user-driven social engagements of news consumption on social media platform. Social engagements represent the news proliferation process over time, which provides useful auxiliary information to infer the veracity of news articles. Generally, there are three major aspects of the social media context that we want to represent: users, generated posts, and networks. First, fake news pieces are likely to be created and spread by non-human accounts, such as social bots or cyborgs [126]. Thus, capturing users' profiles and behaviors by user-based features can provide useful information for fake news detection. Second, people express their emotions or opinions toward fake news through social media posts, such as skeptical opinions and sensational reactions. Thus, it is reasonable to extract post-based features to help find potential fake news via reactions from the general public as expressed in posts. Third, users form different types of networks on social media in terms of interests, topics, and relations. As we mentioned in Section 1.3.1, fake news dissemination processes tend to form an echo chamber cycle, highlighting the value of extracting network-based features to detect fake news.

1.3.5   CHALLENGING PROBLEMS OF FAKE NEWS DETECTION

As in Section 1.3.2, the standard fake news detection problem can be formalized as a classification task. Recent advancements of machine learning methods, such as deep neural networks, tensor factorization, and probabilistic models, allow us to better capture effective features of news from its auxiliary information. In addition, in the real-world scenarios, we face problems. For example, detecting fake news in the early stage is important to prevent its further propagation on social media. As another example, since obtaining the ground truth of fake news is labor-intensive and time-consuming, it is important to study fake news detection in a weakly supervised setting, i.e., with limited or even no labels. Moreover, it is necessary to understand why a particular piece of news is detected as fake by machine learning models, in which the derived explanation can provide new insights, knowledge, and justification non-obvious to practitioners.

Organization   The remainder of this book is organized as follows. In Chapter 2, we introduce the means of utilizing news content for fake news detection. We continue to present how social context information can be utilized to help fake news detection in Chapter 3. In Chapter 4, we discuss different unique problems for fake news detection. Finally, we summarize the representative tools and data repositories for fake news detection on social media. In the two appendices, we provide details of data and tools (or software systems) for fake news detection research.

Notations   To help readers better understand the correlations and differences between fake news detection methods, we propose a unified notations throughout the book as in Table 1.1. In general, we use bold uppercase characters to denote matrix (e.g., $\mathbf{A}$), bold lowercase characters to denote vector (e.g., $\mathbf{w}$), (i, j)-th entry of $\mathbf{A}$ as $\mathbf{A}_{ij}$.

Table 1.1: The main notations

Notation	Meaning	
$\mathcal{A} = \{a_1, \ldots, a_N\}$	the set of news articles	
$\mathbf{a}_i$	the embedding vector of news a_i	
$\mathcal{Y} = \{y_1, \ldots, y_N\}$	the set of labels for corresponding news pieces	
$\mathcal{P} = \{p_1, \ldots, p_K\}$	the set of news publishers	
$\mathcal{S} = \{s_1, \ldots, s_L\}$	the set of sentences for a news piece	
$s_i = \{w_{i1}, \ldots, w_{iMi}\}$	the set of words in a sentence s_i	
$\mathbf{W}_i$	the embedding vector for word w_i	
$\mathcal{I} = \{i_1, \ldots, i_M\}$	the set of images for a news piece	
$\mathcal{E} = \{e_{it,}\}$	the set of social engagements	
$\{u_i, c_i, t\}$	each engagement representing that a user u_i spreads news article a using c_i at time t	
$\mathcal{U} = \{u_1, \ldots, u_n\}$	the set of users engaged in the spreading for a news piece	
$\mathbf{u}_i$	the embedding vector of users	
$\mathcal{C} = \{c_1, \ldots, c_n\}$	the set of social media posts in the spreading of a news piece	
$\mathbf{c}_i$	the embedding vector of post c_i	
$c_i = \{w_{i1}, \ldots, w_{iQi}\}$	the set of words in a post c_i	
Σ	the word vocabulary of text corpus	
$\mathbf{X} \in \mathbb{R}^{N \times d}$	the news-word matrix with $d = |\Sigma|$	
$\mathbf{X} \in \mathbb{R}^{N \times d \times d}$	the news-word-word tensor with $d = |\Sigma|$	
$\mathbf{Y} \in \mathbb{R}^{N \times m \times l}$	the news-user-community tensor	
(s, p, o)	the tuple representing a knowledge: a subject s is related to the object entity by the predicate relation p	
$\lambda(t)$	the intensity function of temporal point process	
$\mathbb{E}(\cdot)$	the empirical expectation function	
$Beta(\alpha, \beta)$	the beta distribution with hyperparameters α and β	
$\mathbf{A} \in \mathbb{R}^{m \times m}$	the user-user adjacency matrix among users engaged in news spreading	
$\mathbf{U} \in \mathbb{R}^{m \times k}$	the user latent representation matrix with latent dimension k	
$\mathbf{E} \in \mathbb{R}^{m \times N}$	the user-news interaction matrix	
$\mathbf{F} \in \mathbb{R}^{N \times k}$	the news latent representation matrix with latent dimension k	

CHAPTER 2

What News Content Tells

News content features describe the meta information related to a piece of news. A list of representative news content features are listed as follows.

- **Source:** Author or publisher of the news article.

- **Headline:** Short title text that aims to catch the attention of readers and describes the main topic of the article.

- **Body Text:** Main text that elaborates the details of the news story; there is usually a major claim that is specifically highlighted and that shapes the angle of the publisher.

- **Image/Video:** Part of the body content of a news article that provides visual cues to frame the story.

Based on these raw content features, different kinds of feature representations can be built to extract discriminative characteristics of fake news. Typically, the news content we are looking at will mostly be *textual features*, *visual features*, and *style features*.

2.1 TEXTUAL FEATURES

Textual features are extracted from news content with natural language processing (NLP) techniques [102]. Next, we introduce how to extract linguistic, low-rank, and neural textual features.

2.1.1 LINGUISTIC FEATURES

Linguistic features are extracted from the text content in terms of characters, words, sentences, and documents. In order to capture the different aspects of fake news and real news, both common linguistic features and domain-specific linguistic features are utilized. Common linguistic features are often used to represent documents for various tasks in natural language processing. Common linguistic features are: (i) *lexical features*, including character-level and word-level features, such as total words, characters per word, frequency of large words, and unique words; and (ii) *syntactic features*, including sentence-level features, such as frequency of function words and phrases (i.e., "n-grams" and bag-of-words approaches [43]) or punctuation and parts-of-speech (POS) tagging [5, 107]. Domain-specific linguistic features, which are specifically aligned to news domain, such as quoted words, external links, number of graphs, and the average length of graphs [111].

2.1.2   LOW-RANK TEXTUAL FEATURES

Low-rank modeling for text data has been widely explored in different domains [155]. Low-rank approximation is to learn a compact (low-dimension) text representations from the high-dimension and noisy raw feature matrix. Existing low-rank models are mostly based on matrix factorization or tensor factorization techniques, which project the term-news matrix into a k-dimensional latent space.

Matrix Factorization

Matrix factorization for text modeling has been widely used for learning document representations such as clustering [173]. It can learn a low-dimensional basis where each dimension represents the coherent latent topic from raw feature matrix. Non-negative matrix factorization (NMF) methods introduce non-negative constraints when factorizing the news-words matrix [138]. Using NMF, we attempt to project the document-word matrix to a joint latent semantic factor space with low dimensionality, such that the document-word relations are modeled as the inner product in the space. Specifically, giving the news-word matrix $\mathbf{X} \in \mathbb{R}_+^{n \times d}$, NMF methods try to find two non-negative matrices $\mathbf{F} \in \mathbb{R}_+^{N \times k}$ and $\mathbf{V} \in \mathbb{R}_+^{d \times k}$ by solving the following optimization problem:

$$\min_{\mathbf{F}, \mathbf{V} \geq 0} \left\| \mathbf{X} - \mathbf{F}\mathbf{V}^T \right\|_F^2 , \tag{2.1}$$

where d is the size of word vocabulary and k is the dimension of the latent topic space. In addition, $\mathbf{R}$ and $\mathbf{V}$ are the non-negative matrices indicating low-dimensional representations of news and words.

Tensor Factorization

The goal of tensor decomposition is to learn the representation of news articles by considering the spatial relations of words in a document. Tensor factorization approaches [46, 53] first build a 3-mode news-word-word tensor for a news document as $\underline{\mathbf{X}} \in \mathbb{R}^{N \times d \times d}$, where a horizontal slice $\underline{\mathbf{X}}_{i,:,:}$ represents the spatial relations in a news document; and for a horizontal slice $\mathbf{S}$, $\mathbf{S}(i, j)$ represents the number of times that the ith term and the jth term of the dictionary appear in an affinity of each other. To learn the representations, different tensor decomposition techniques can be applied. For example, we can use nonnegative CP/PARAFAC decomposition with alternating Poisson regression (CP APR) that uses Kullback–Leibler divergence because of very high sparsity in the tensor. CP/PARAFAC decomposition represents $\underline{\mathbf{X}}$ as follows:

$$\underline{\mathbf{X}} \approx [\mathbf{F}, \mathbf{B}, \mathbf{H}] = \sum_{r=1}^{R} \lambda_r \mathbf{f}_r \odot \mathbf{b}_r \odot \mathbf{h}_r, \tag{2.2}$$

where $\odot$ denotes the outer product and $\mathbf{f}_r$ (same for $\mathbf{b}_r$ and $\mathbf{h}_r$) denotes the normalized rth column of non-negative factor matrix $\mathbf{F}$ (same for $\mathbf{B}$ and $\mathbf{H}$), and R is the rank. Each row of $\mathbf{F}$ denotes the representation of the corresponding article in the embedding space.

2.1.3   NEURAL TEXTUAL FEATURES

With the recent advancements of deep neural networks in NLP, different neural network structures, such as convolution neural networks (CNNs) and recurrent neural networks (RNNs), are developed to learn the latent textual feature representations. Neural textual features are based on dense vector representations rather than high-dimensional and sparse features, and have achieved superior results on various NLP tasks [178]. We introduce major representative deep neural textual methods including CNNs and RNNs, and some of their variants.

CNN

CNNs have been widely used for fake news detection and achieve many good results [163, 175]. CNNs have the ability to extract salient n-gram features from the input sentence to create an informative latent semantic representation of the sentence for downstream tasks [178]. As shown in Figure 2.1, to learn the textual representations of news sentences, CNNs first build a word representation matrix for each word using the word-embedding vectors such as Word2Vec [90], then apply several convolution layers and a max-pooling layer to obtain the final textual representations. Specifically, for each word w_i in the news sentence, we can denote the k dimensional word embedding vector as $\mathbf{w}_i \in \mathbb{R}^k$, and a sentence with n words can be represented as:

$$\mathbf{w}_{1:n} = \mathbf{w}_1 \oplus \mathbf{w}_2 \cdots \mathbf{w}_n, \tag{2.3}$$

where $\oplus$ denotes the concatenation operation. A convolution filter with window size h takes the contiguous sequence of h words in the sentence as input and output the feature, as follows:

$$\tilde{\mathbf{w}}_i = \sigma(\mathbf{W} \cdot \mathbf{w}_{i:i+h-1}) \tag{2.4}$$

and $\sigma(\cdot)$ is the ReLU activation function and $\mathbf{W}$ represents the weight of the filter. The filter can further be applied to the rest of the words and then we can get a feature vector for the sentence:

$$\tilde{\mathbf{w}} = [\tilde{\mathbf{w}}_1, \tilde{\mathbf{w}}_2, \cdots, \tilde{\mathbf{w}}_{n-h+1}] \tag{2.5}$$

for every feature vector $\mathbf{t}$, we further use max-pooling operation to take the maximum value to extract the most important information. The process can be repeated until we get the features for all filters. Following the max pooling operations, a fully connected layer is used to ensure the final textual feature representation.

RNN

An RNN is popular in NLP, which can encode the sequence information of sentences and paragraphs directly. The representative RNN for learning textual representation is the long short-term memory (LSTM) neural networks [64, 114, 119].

For example, two layers of LSTM [114] is built to detect fake news, where one layer puts simple word embedding into LSTM and the other one concatenate LSTM output with Linguistic Inquiry and Word Count (LIWC) [105] feature vectors before feeding into the action layer.

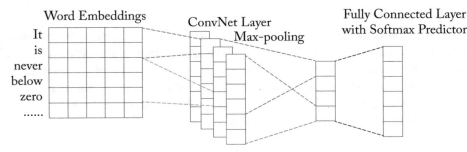

Figure 2.1: The illustration of CNNs for learning neural textual features. Based on [163].

In both cases, they were more accurate than NBC and Maximum Entropy models. Karimi and Tang [64] utilized a bi-directional LSTM to learn the sentence-level textual representations. To encode the uncertainty of news contents, Zhang et al. [182] proposed to incorporate a Bayesian LSTM to learn the representations of news claims. Recent works [69, 81] incorporated speakers' profile information such as names and topics to LSTM model to predict fake news. Moreover, Pham et al. utilized memory networks, which is a kind of attention-based neural network, to learn textual representations by memorizing a set of words in the memory [109].

In addition to modeling the sentence-level content with RNNs, other approaches model the news-level content in a hierarchical way such as hierarchical attention neural networks [177] and hierarchical sequence to sequence auto-encoders [78]. For hierarchical attention neural networks (see Figure 2.2), we first learn the sentence vectors by using the word encoder with attention and then learn the sentence representations through sentence encoder component. Specifically, we can use bidirectional gated recurrent units (GRU) [8] to model word sequences from both directions of words:

$$\overrightarrow{\mathbf{h}_{it}} = \overrightarrow{GRU}(\mathbf{w}_{it}), t \in \{1, \ldots, M_i\}$$
$$\overleftarrow{\mathbf{h}_{it}} = \overleftarrow{GRU}(\mathbf{w}_{it}), t \in \{1, \ldots, M_i\}. \tag{2.6}$$

We then obtain an annotation of word w_{it} by concatenating the forward hidden state $\overrightarrow{\mathbf{h}_{it}}$ and backward hidden state $\overleftarrow{\mathbf{h}_{it}}$, i.e., $\mathbf{h}_{it} = [\overrightarrow{\mathbf{h}_{it}}, \overleftarrow{\mathbf{h}_{it}}]$, which contains the information of the whole sentence centered around w_i. Note that not all words contribute equally to the representation of the sentence meaning. Therefore, we introduce an attention mechanism to learn the weights to measure the importance of each word, and the sentence vector $\mathbf{v}_i \in \mathbb{R}^{2d \times 1}$ is computed as follows:

$$\mathbf{v}_i = \sum_{t=1}^{M_i} \alpha_{it}\mathbf{h}_{it}, \tag{2.7}$$

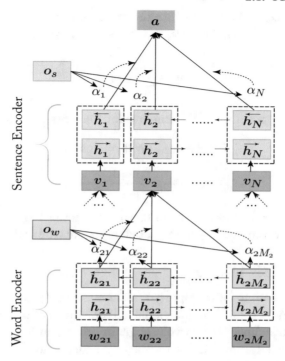

Figure 2.2: The illustration of hierarchical attention neural networks for learning neural textual features. Based on [163].

where α_{it} measures the importance of tth word for the sentence s_i, and α_{it} is calculated as follows:

$$\mathbf{o}_{it} = \tanh(\mathbf{W}_w \mathbf{h}_{it} + \mathbf{b}_w)$$

$$\alpha_{it} = \frac{\exp(\mathbf{o}_{it}\mathbf{o}_w^\mathsf{T})}{\sum_{k=1}^{M_i} \exp(\mathbf{o}_{ik}\mathbf{o}_w^\mathsf{T})}, \qquad (2.8)$$

where $\mathbf{o}_{it}$ is a hidden representation of $\mathbf{h}_{it}$ obtained by feeding the hidden state $\mathbf{h}_{it}$ to a fully embedding layer, and $\mathbf{o}_w$ is the weight parameter that represents the world-level context vector. Similarly, RNNs with GRU units to encode each sentence in news articles,

$$\overrightarrow{\mathbf{h}_i} = \overrightarrow{GRU}(\mathbf{v}_i), i \in \{1, \ldots, N\}$$

$$\overleftarrow{\mathbf{h}_i} = \overleftarrow{GRU}(\mathbf{v}_i), i \in \{N, \ldots, 1\}. \qquad (2.9)$$

We then obtain an annotation of sentence $\mathbf{h}_i \in \mathbb{R}^{2d \times 1}$ by concatenating the forward hidden state $\overrightarrow{\mathbf{h}_i}$ and backward hidden state $\overleftarrow{\mathbf{h}_i}$, i.e., $\mathbf{h}_i = [\overrightarrow{\mathbf{h}_i}, \overleftarrow{\mathbf{h}_i}]$, which captures the context from neighbor sentences around sentence s_i. We can also introduce an attention mechanism to learn the weights to measure the importance of each sentence, and the news article vector $\mathbf{a}$ is computed as follows:

$$\mathbf{a} = \sum_{i=1}^{N} \alpha_i \mathbf{h}_i, \tag{2.10}$$

where α^i measures the importance of ith sentence for the news piece a, and α_i is calculated as follows:

$$\mathbf{o}_i = \tanh(\mathbf{W}_s \mathbf{h}_i + \mathbf{b}_s)$$
$$\alpha_i = \frac{\exp(\mathbf{o}_i \mathbf{o}_s^\mathsf{T})}{\sum_{k=1}^{N} \exp(\mathbf{o}_k \mathbf{o}_s^\mathsf{T})}, \tag{2.11}$$

where $\mathbf{o}_i$ is a hidden representation of $\mathbf{h}_i$ obtained by feeding the hidden state $\mathbf{h}_i$ to a fully embedding layer, and $\mathbf{o}_s$ is the weight parameter that represents the sentence-level context vector.

2.2 VISUAL FEATURES

Visual cues have been shown to be an important manipulator for fake news propaganda.[1] As we have described, fake news exploits the individual vulnerabilities of people and thus often relies on sensational or even fake images to provoke anger or other emotional response of consumers. Visual features are extracted from visual elements (e.g., images and videos) to capture the different characteristics for fake news. Visual features are generally categorized into three types [21]: *Visual Statistical Features*, *Visual Content Features*, and *Neural Visual Features*.

2.2.1 VISUAL STATISTICAL FEATURES

Visual statistical features represent the statistics attached to fake/real news pieces. Some representative visual statistical features include [60] the following.

- **Count**: the occurrence of images in fake news pieces, they count the total images in a news event and the ratio of news posts containing at least one or more than one images.

- **Popularity**: the popularity of images indicate the number of sharing on social media.

- **Image type**: some images have particular type in resolution or style. For example, long images are images with a very large length-to-width ratio. The ratio of these types of images is also counted as a statistical feature.

2.2.2 VISUAL CONTENT FEATURES

Research [60] has shown that image contents in fake news and real news have different characteristics. The representative visual content features are detailed as follows.

---

[1]https://www.wired.com/2016/12/photos-fuel-spread-fake-news/

- **Visual Clarity Score (VCS)**: Measures the distribution difference between two image sets: one is the image set in a certain news event (event set) and the other is the image set containing images from all events (collection set). Visual clarity score is measured as the Kullback–Leibler divergence between two language models representing the event image set and all image set, respectively. The bag-of-word image representation, such as SIFT [82] or SURF [11] features, can be used to define language models for images. Specifically, let $p(w|c)$ and $p(w|k)$ denote the term frequency of visual word w in collection set and event set, respectively, and the visual clarity score is denoted as

$$VCS = D_{KL}(p(w|c)\|p(w|k)). \tag{2.12}$$

- **Visual Coherence Score (VCoS)**: Measures the coherence degree of images in a certain news event. This feature is computed based on visual similarity among images and can reflect relevant relations of images in news events quantitatively. More specifically, the average of similarity scores between every two images i_j and i_k are computed as the coherence score as follows:

$$VCoS = \frac{1}{|M(M-1)|} \sum_{j,k=1,\cdots,M; j \neq k} sim(i_i, i_k). \tag{2.13}$$

Here M is number of images in event set and $sim(i_j, i_k)$ is the visual similarity between image i_j and image i_k. In implementation, the similarity between the image pairs is calculated based on their GIST [101] feature representations.

- **Visual Similarity Distribution Histogram (VSDH)**: Describes inter-image similarity in a fine granularity level. It evaluates image distribution with a set of values by quantifying the similarity matrix of each image pair in an event. The visual similarity matrix $\mathbf{S}$ is obtained by calculating pairwise image similarity in a news event. The visual similarity is also computed based on their GIST [101] feature representations. The similarity matrix $\mathbf{S}$ is then quantified into an H-bin histogram by mapping each element in the matrix into its corresponding bin, which results in a feature vector of H dimensions representing the similarity relations among images,

$$VSDH(h) = \frac{1}{M^2} |\{(j,k)|j,k \leq M, m_{j,k} \in h_{th}bin\}|, h = 1, \cdots, H. \tag{2.14}$$

- **Visual Diversity Score (VDS)**: Measures the visual difference of the image distribution. First, images are ranked according to their popularity on social media, based on the assumption that popular images would have better representation for the news event. Then, the diversity of an image is defined as its minimal difference with the images ranking before it in the entire image set [161]. At last, the visual diversity score is then calculated as a weighted average of dissimilarity over all images, where top-ranked images have larger weights [35],

$$VDS = \sum_{j=1}^{M} \frac{1}{j} \sum_{k=1}^{j} (1 - sim(i_j, i_k)). \qquad (2.15)$$

• **Visual Clustering Score**: Evaluates the image distribution over all images in the news event from a clustering perspective. Representative clustering methods such as hierarchical agglomerative clustering [66] (HAC) algorithm can be utilized to obtain the image clusters.

2.2.3   NEURAL VISUAL FEATURES

Multi-layer neural networks have been widely used for learning image feature representations. Specifically, the specially designed architecture of CNNs are very powerful in extracting visual features from images, which can be used for various tasks [143, 162]. VGG 16 is one the state-of-the-art CNNs (see Figure 2.3) for learning neural visual representations [143]. It is comprised of three basic types of layers; convolutional layers for extracting translation-invariant features from images, pooling layers for reducing the parameters, and fully connected layers for classification tasks. To prevent CNN from over-fitting and to ease the training of deep CNNs, dropout layers [145] and residual layers [50] are introduced to CNN structures. Recent work that use images for fake news detection has adopted the VGG model [57, 143] to extract neural visual features [165].

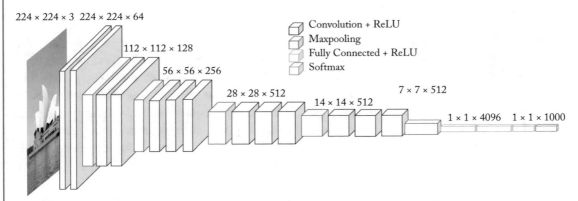

Figure 2.3: The illustration of VGG 16 framework for learning neural image features.

2.3   STYLE FEATURES

Fake news publishers often have malicious intent to spread distorted and misleading information and influence large communities of consumers, requiring particular writing styles necessary to appeal and persuade a wide scope of consumers that is not seen in true news articles. Style approaches try to detect fake news by capturing the *manipulators* in the writing style of the news content. There are mainly three typical categories of style-based methods: *Deception Styles*, *Clickbaity Styles*, and *New Quality Styles*.

2.3.1   DECEPTION STYLES

The motivation of deception detection originates from forensic psychology (i.e., Undeutsch Hypothesis) [156] and various forensic tools including Criteria-based Content Analysis [160] and Scientific-based Content Analysis [77] have been developed. More recently, advanced natural language processing models are applied to spot deception phases from the following perspectives: *Deep syntax* and *Rhetorical structure*. Deep syntax models have been implemented using probabilistic context free grammars (PCFG), with which sentences can be transformed into rules that describe the syntax structure. Based on the PCFG, different rules can be developed for deception detection, such as unlexicalized/lexicalized production rules and grandparent rules [40]. Rhetorical structure theory can be utilized to capture the differences between deceptive and truthful sentences [118]. Moreover, other features can be specifically designed to capture the deceptive cues in writing styles to differentiate fake news, such as lying-detection features [2].

2.3.2   CLICKBAITY STYLES

Since fake news pieces are intentionally created for financial or political gain rather than for objective claims, they often contain opinionated and inflammatory language, crafted as "clickbait" (i.e., to entice users to click on the link to read the full article) or to incite confusion [25]. Thus, it is reasonable to exploit linguistic features that capture the different writing styles and sensational headlines to detect fake news [137]. Biyani et al. [16] studied the characteristics of page "clickbaits," whose news headlines were more interesting or appealing than the actual article. We introduce the following Clickbaity Style features.

Content: Content features are used to quantify the certain content type and formatting such as superlative (adjectives and adverbs), quotes, exclamations, use of uppercase letters, asking questions, etc. Table 2.1 shows the content features.

Informality: Fake news as well as clickbaits can often be sensational, provoking, and gossip-like content. Therefore, their language tends to be less formal than that of professionally written news articles. Thus, Biyani et al. use the following scores to indicate the readability/informality level of a text.

- **Coleman–Liau score (CLScore):** computed as $0.0588L - 0.296S - 15.8$ where L is the average number of letters and S is the average number of sentences per 100 words.

Table 2.1: Description of content features. Feature type "N" and "B" imply that **numeric** and **binary**, respectively.

Feature	Description	Type
NumWords	Number of words	
NumCap	Number of uppercase words (excluding acronyms: words with less than five characters)	N
NumAcronym	Number of acronyms (uppercase words with less than fve characters)	N
Is,NumExclm	Presence/Number of exclamation marks	B/N
Is,NumQues	Presence/Number of question marks	B/N
IsStartNum HasNumber	Whether the title starts with a number	B
HasNumber	Whether the title contains a number (set only if the title doesn't start with a number)	B
IsSuperlative	Presence of superlative adverbs and adjectives (POS tags RBS, JJS)	B
Is,NumQuote	Presence/Number of quoted words (used "','; excluded 'm, 're, 've, 'd, 's, s')	B/N
IsStart5W1H	Whether the title starts with 5W1H words (what, why, when, who, which, how)	B
Is,NumNeg	Presence/Number of negative sentiment words	B/N

- **RIX and LIX indices** (Anderson 1983 [3]): computed as $RIX = LW/S$ and $LIX = W/S + (100LW)/W$ where W is the number of words, LW is the number of long words (7 or more characters), and S is the number of sentences.

- **Formality measure (fmeasure)** (Heylighen and Dewaele 1999 [51]): the score is used to calculate the degree of formality of a text by measuring the amount of different part-of-speech tags in it. It is computed as $(nounfreq + adjectivefreq + prepositionfreq + articlefreq - pronounfreq - verbfreq - adverbfreq - interjectionfreq + 100)/2$.

The combination of content and informality features are then formalized as the clickbaity style features.

2.3.3 NEWS QUALITY STYLES

News quality is a comprehensive indicator used to measure the readability, the amount of information, and the writing formality of news. Real news that is published by a professional journalist or news agency is usually with high news quality, however the fake news that aimed at misleading readers tends to have poor news quality.

Yang et al. [176] proposed eight types of linguistic features based on news writing guidelines to represent writing style, and studied the correlation between the writing style and news quality. They conducted research on eight aspects: readability, credibility, interactivity, sensation, logic, formality, interestingness, and structural integrity (see Table 2.2).

Table 2.2: Description of news quality style features

Categories	Features	Description
Readability	Sentence_broken, Characters, Words, Sentences, Clauses, Average_word_length, Professional_words, RIX, LIX, LW	Measuring the clarity and legibility of the news
Credibility	#@, Numerals, Ocial_speech, Time, Place, Object, Uncertainty, Image	Measuring the rigor and reliability of the news
Interactivity	Question_mark, First_pron, Second_pron, Interrogative_pron	Measure the interactivity between the news and the reader
Sensation	Sentiment_score, Adv_of_degree, Modal_particle, First_pronoun, Second_pronoun, Exclamation_mark, Question_mark	Measure the impression that the news leaves on the reader
Logic	Forward reference, Conjunctions	Determining whether the news is logical and contextually coherent or not
Formality	Noun, Adj, Prep, Pron, Verb, Adv, Sentence_broken	Measuring the formality of news language
Interestingness	Rhetoric, Exclamation mark, Face, Idiom, Adv, Adj	Measuring the interestingness of news language
Structural Integrity	HasHead, HasImage, HasVideo, HasTag, HasAt, HasUrl	Measuring the structural integrity of news

2.4 KNOWLEDGE-BASED METHODS

Since fake news attempts to spread false claims in news content, one of the most straightforward means of detecting it is to check the truthfulness of major claims in a news article to decide the news veracity. Knowledge-based approaches use sources that are employed to fact-check claims in news contents. The goal of fact-checking is to assign a truth value to a claim in a particular context [158]. Fact-checking has attracted increasing attention, and many efforts have been made to develop an automated fact-checking system. The goal is to assess news authenticity by comparing the information extracted from to-be-verified news content with known knowledge. Existing fact-checking approaches can be categorized as *Manual Fact-checking* and *Automatic Fact-checking*.

2.4.1   MANUAL FACT-CHECKING

Manual fact-checking aims to utilize human experts to provide signals manually of annotating fake news. It heavily relies on human domain experts or normal users to investigate relevant data and documents to construct the verdicts of claim veracity. Existing manual fact-checking approaches mainly fall into: *expert-based* and *crowdsourcing-based* fact-checking (Table 2.3).

Table 2.3: Comparison of expert-based and crowdsourcing-based fact checking

	Expert-Based	Crowdsourcing-Based
Fact-checkers	Domain-experts	Regular individuals (i.e., collective intelligence)
Annotation reliability	High	Comparatively low
Scalability	Poor	Comparatively high

Expert-Based Fact-Checking

Fact-checking heavily relies on human domain experts to investigate relevant data and documents to deliver the verdicts of claim veracity. However, expert-based fact-checking is an intellectually demanding and time-consuming process, which limits the potential for high efficiency. We introduce some representative and popular fact-checking websites as follows.

- **PolitiFact**:[2] PolitiFact is a U.S. website that rates the accuracy of claims or statements by elected officials, pundits, columnists, bloggers, political analysts, and other members of the media. It is an independent, non-partisan source of online fact-checking system for political news and information. The editors examine the specific word and the full context of a claim carefully, and then verify the reliability of the claims and statements. The label types include true, mostly true, half true, mostly false, false, and pants on fire.

- **Snopes**:[3] Snopes is widely known as one of the first online fact-checking websites for validating and debunking urban legends. It covers a wide range of disciplines including

[2]www.politifact.com/
[3]http://www.snopes.com/

automobiles, business, computers, crime, fraud and scams, history, and so on. The label types include true and false.

- **FactCheck**:[4] FactCheck is a nonprofit "consumer advocate webpage" for voters that aims to reduce the level of deception and confusion in U.S. politics. Those claims and statements are originated from various platforms, including TV advertisements, debates, speeches, interviews, new releases, and social media. They mainly focus on presidential candidates in presidential election years, and evaluate the factual accuracy of their statements systematically.

- **GossipCop**:[5] GossipCop investigates entertainment stories that are published in magazines and newspapers, as well as on the Web, to ascertain whether they are true or false. They provide the score scaling from 0–10, where 0 means fake and 10 mean real.

- **TruthOrFiction**:[6] TruthOrFiction is a non-partisan online website that provide fact-checking results on warnings, hoaxes, virus warnings, and humorous or inspirational stories that are distributed through emails. It mainly focuses on misleading information that are popular via forwarded emails. And they rate stories or information by the following categories: truth, fiction, reported to be truth, unproven, truth and fiction, previously truth, disputed, and pending investigation.

Crowdsourcing-Based Fact-Checking

Fact-checking exploits the "wisdom of crowd" to enable people to annotate news content. These annotations are then aggregated to produce an overall assessment of the claim veracity. For example, Fiskkit[7] allows users to discuss and annotate the accuracy of specific parts of a news article. As another example, an anti-fake-news bot named "For real" is a public account in the communication mobile application LINE,[8] which allows people to report suspicious news content which is then further checked by editors.

2.4.2   AUTOMATIC FACT-CHECKING

Manual fact-checking relies on humans annotation, which is usually time-consuming and labor-intensive. Instead, automatic fact-checking for specific claims largely relies on *external knowledge* to determine the truthfulness of a particular claim. Two typical external sources include the *open web* and structured *knowledge graph*. Open web sources are utilized as references that can be compared with given claims in terms of both the consistency and frequency [10, 84]. Knowledge graphs are integrated from the linked open data as a structured network topology, such as

---

[4]https://www.factcheck.org/
[5]https://www.gossipcop.com/
[6]https://www.truthorfiction.com/
[7]http://fiskkit.com
[8]https://grants.g0v.tw/projects/588fa7b382223f001e022944

DBpedia and Google Relation Extraction Corpus. Fact-checking using a knowledge graph aims to check whether the claims in news content can be inferred from existing facts in the knowledge graph [29, 129, 171]. Next, we introduce a standard knowledge graph matching approach that matches news claims with the facts in knowledge graphs.

Path Finding   Fake news spreads false claims in news content, so a natural means of detecting fake news is to check the truthfulness of major claims in the news article. A claim in news content can be represented by a subject-predicate-object triple (s, p, o), where the subject entity s is related to the object entity o by the predicate relation p. We can find all the paths that start with s and end with o, and then evaluate these paths to estimate the truth value of the claim. This set of paths, also known as knowledge stream [130], are denoted as $\mathcal{P}(s, o)$. Intuitively, if the paths involve more specific entities, then the claim is more likely to be true. Thus, we can define a "specificity" measure $S(P_{s,o})$ as follows:

$$S(P_{s,o}) = \frac{1}{1 + \sum_{i=2}^{n-1} \log d(o_i)}, \tag{2.16}$$

where $d(o_i)$ is the degree of entity o_i, i.e., the number of paths that entity o participates. One approach is to optimize a path evaluation function: $\tau(c) = \max \mathcal{W}(P_{s,o})$, which maps the set of possible paths connecting s and o (i.e., $P_{s,o}$) to a truth value τ. If s is already present in the knowledge graph, it can assign maximum truth value 1; otherwise, the objective function will be optimized to find the shortest path between s and o.

Flow Optimization   We can assume that each edge of the network is associated with two quantities: a *capacity* to carry knowledge related to (s, p, o) across its two endpoints, and a *cost* of usage. The capacity can be computed using $S(P_{s,o})$, and the cost of an edge in knowledge is defined as $c_e = \log d(o_i)$. The goal is to identify the set of paths responsible for the maximum flow of knowledge between s and o at the minimum cost. The maximum knowledge a path $P_{s,o}$ can carry is the minimum knowledge of its edges, also called its bottleneck $B(P_{s,o})$. Thus, the objective can be defined as a minimum cost maximum flow problem

$$\tau(e) = \sum_{P_{s,o} \in \mathcal{P}_{s,o}} B(P_{s,o}) \cdot S(P_{s,o}), \tag{2.17}$$

where $B(P_{s,o})$ is denoted as a minimization form: $B(P_{s,o}) = \min\{x_e | \in P_{s,o}\}$, with x_e indicating the residual capacity of edge x in a residual network [130].

The knowledge graph itself can be redundant, invalid, conflicting, unreliable, and incomplete [185]. In these cases, path finding and flow optimization may not be sufficient to obtain good results of assessing the truth value. Therefore, additional tasks need to be considered in order to reconstruct the knowledge graph and to facilitate its capability as follows.

- **Entity Resolution:** refers to the process of finding related entries in one or more related relations in a database and creating links among them [19]. This problem has been extensively studied in the database area and applied to data warehousing and business intelligence. Based on this survey [72], existing methods exploit features in three ways, namely numerical, rule-based, and workflow-based. Numerical approaches combine the similarity score of each feature into a weighted sum to decide linkage [39]; rule-based approaches derive match decision through a logical combination of testing separate rules of each feature with a threshold; workflow-based methods apply a sequence of feature comparison in an iterative way. Both supervised such as TAILOR [37] and MARLIN [15], and unsupervised approaches such as MOMA [151] and SERF [13] are studied in the literature.

- **Time Recording:** aims to remove outdated knowledge. This task is important giving that fake news pieces are often related to newly emerging events. Existing work on time recording mainly utilize the Compound Value Type structure to allow facts incorporating beginning and ending date annotations [17], or adding extra assertions to current facts [52].

- **Knowledge Fusion:** (or truth discovery) aims to identify true subject-predicate-object triples extracted by multiple information extractors from multiple information sources [36, 79]. Truth discovery methods do not explore the claims directly, but rely on a collection of contradicting sources that record the properties of objects to determine the truth value. Truth discovery aims to determine the *source credibility* and *object truthfulness* at the same time. Fake news detection can benefit from various aspects of truth discovery approaches under different scenarios. For example, the credibility of different news outlets can be modeled to infer the truthfulness of reported news. As another example, relevant social media posts can also be modeled as social response sources to better determine the truthfulness of claims [93, 167]. However, there are some other issues that must be considered to apply truth discovery to fake news detection in social media scenarios. First, most existing truth discovery methods focus on handling *structured* input in the form of subject-predicate-object (SPO) tuples, while social media data is highly unstructured and noisy. Second, truth discovery methods cannot be well applied when a fake news article is newly launched and published by only a few news outlets because at that point there is not enough social media posts relevant to it to serve as additional sources.

- **Link Prediction:** on knowledge graphs aims to predict new fact from existing facts. This is important since existing knowledge graphs are often missing many facts, and some of the edges they contain are incorrect. Relational machine learning methods are widely used to infer new knowledge representations [97], including latent feature models and graph feature models. Latent feature models exploit the latent features or entities to learn the possible SPO triples. For example, RESCAL [98] is a bilinear relational learning model that explain triples through pairwise interactions of latent features. Graph feature models assume that the existence of an edge can be predicted by extracting features from the observed edges in the graph, such as Markov logic programming or path ranking algorithms. For example, Markov Random Fields (MRFs) [129] encode dependencies of facts into random variables and infer the missing dependencies through statistical probabilistic learning.

CHAPTER 3

How Social Context Helps

Social context refers to the entire social environment in which the dissemination of the news operates, including how the social data is distributed and how online users interact with each other. It provides useful auxiliary information for inferring the veracity of news articles. The nature of social media provides researchers with additional resources to supplement and enhance news content-based models. There are three major aspects of the social context information that we can represent: users, generated posts, and networks. First, users may have different characteristics for those spreading fake and real news, or establish different patterns of behaviors toward fake news. Second, in the process of fake news dissemination, users express their opinions and emotions through posts/comments. Third, users form different types of networks on social media. We now discuss how social context information can help fake news detection from three perspectives: *user-based*, *post-based*, and *network-based*.

3.1 USER-BASED DETECTION

User-based fake news detection aims to explore the characteristics and behaviors of consumers on social media to classify fake news. Next, we first demonstrate the comparison and exploration of user profiles for fake news detection, and then discuss how to model user behaviors such as "flagging fake news" to predict fake news.

3.1.1 USER FEATURE MODELING

Previous research advancements aggregate uses profiles and engagements on news pieces to help infer which articles are fake [22, 59], giving some promising early results. Recent research starts to perform a principled study on characterizing user profiles and explore their potential to detect fake news.

Profile Features

We collect and analyze user meta profile features from two major aspects, i.e., *explicit* and *implicit* [142]. Explicit features are obtained directly from meta-data returned by querying social media site aplication programming interface (APIs). Implicit features are not directly available but are inferred from user meta information or online behaviors, such as historical tweets. Our selected feature sets are by no means the comprehensive list of all possible features. However, we focus on those explicit features that can be easily accessed and are available for almost all public users, and implicit features that are widely used in the literature for better understanding user characteristics. We first select two subset of users who are more likely to share fake and real news based on FakeNewsNet data [134], and compare the aggregated statistics over these two sets [139].

Explicit Features   A list of representative explicit profile attributes include the following.

- **Profile-Related.** Basic user description fields:

    - *Verified*: whether this is a verified user;
    - *RegisterTime*: the number of days since the accounted was registered;

- **Content-Related.** Attributes of user activities:

    - *StatusCount*: the number of posts;
    - *FavorCount*: the number of favorites;

- **Network-Related.** Social networks attributes:

    - *FollowerCount*: the number of followers;
    - *FollowingCount*: the number of users being followed.

Implicit Features   We also explore several implicit profile features, which are not directly provided through user meta data, but are widely used to describe and understand user demographics [122]. Note that we adopt widely used tools to predict these implicit features in an *unsupervised* way. Some representative features are as follows.

- **Age:** studies have shown that age has major impacts on people's psychology and cognition. For example, as age gradually changes, people typically become less open to experiences, but more agreeable and conscientious [86]. We infer the age of users using existing state-of-the-art approaches [121]. This method uses a linear regression model with the collected predictive lexica (with words and weights).

- **Personality:** personality refers to the traits and characteristics that makes an individual different from others. We draw on the popular Five Factor Model (or "Big Five"), which classifies personality traits into five dimensions: **E**xtraversion (e.g., outgoing, talkative, active); **A**greeableness (e.g., trusting, kind, generous); **C**onscientiousness (e.g., self-controlled, responsible, thorough); **N**euroticism (e.g., anxious, depressive, touchy); and **O**penness (e.g., intellectual, artistic, insightful). To predict users' personalities, we apply an unsupervised personality prediction tool called Pear [23], a state-of-the-art unsupervised, text-based personality prediction model.

- **Political Bias:** political bias plays an important role in shaping users' profiles and affecting news consumption choices on social media. Sociological studies on journalism demonstrate the correlation between partisan bias and news content authenticity (i.e., fake or real news) [44]. Specifically, users tend to share news that confirms their existing political bias [99]. We adopt a method in [74] to measure user political bias scores by exploiting users' interests.

From empirical comparison analysis, we observe that most of the explicit and implicit profile features reveal different feature distributions, which demonstrates the potential to use them to detect fake news. For example, as shown in Figure 3.1, we demonstrate the box-and-whisker diagram, which shows that the distribution of user RegisterTime exhibits a significant difference between users who spread fake news and those who spread real news. The observations

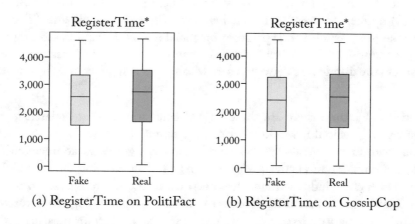

(a) RegisterTime on PolitiFact   (b) RegisterTime on GossipCop

Figure 3.1: **Profile Features Comparison.** We show the Box-Plot to demonstrate the distribution of *RegisterTime* for users. Based on [142].

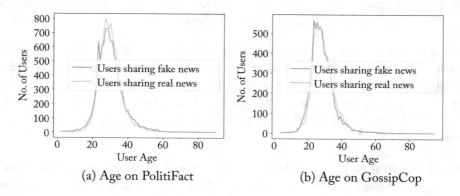

(a) Age on PolitiFact   (b) Age on GossipCop

Figure 3.2: **Age Comparison.** We rank the ages from low to high and plot the values for users. The x-axis represents the predicted ages, and y-axis indicates the number of users. Based on [142].

on both datasets demonstrate that users who are more likely to share fake news registered much earlier. As another example for implicit features, we demonstrate the comparison in Figure 3.2. We can see that the predicted ages are significantly different, and users who spread fake news are predicted younger than those who spread real news. Motivated by the observations, we can further extract these explicit and implicit features for all the users that spread the news to predict whether it is a fake news piece or not.

Psychology-Related Features

To understand the characteristics of users who spread fake news, we can rely on psychological theories. Although there is a large body of work on these psychological theories, not many of them can be (1) applied to users and their behaviors on social media and (2) quantitatively measured for fake news articles and spreaders on social media. Hence, based on psychological theories we have mentioned in Section 1.2, we can enumerate five categories of features that can potentially express the differences between users who spread fake news and the ones who spread real news.

- **Motivational Factors:** there are three LIWC categories that are related to *uncertainty*: discrepancy (e.g., should, would, and could), tentativeness (e.g., maybe, perhaps, and guess), and certainty (e.g., always and never). These categories are abbreviated as `discrep`, `tentat`, and `certain` respectively. *Anxiety* can be measured using the LIWC Anxiety category (`anx`) which includes words such as nervous, afraid, and tense. *Importance or outcome-relevance* is observed to be a difficult feature to measure in psychology so researchers suggest using proxies to quantify *importance*; we use *anxiety* as a proxy for measuring this feature, meaning that people are more anxious about a topic which is more important to them. We use LIWC Future Focus (`futurefocus`) to measure *lack of control*, this category includes words such as may, will, and soon. We do not measure *belief* explicitly because we assume that any user who tweets fake news articles believes in it.

- **Demographics:** Twitter users are not obligated to include information such as age, race, gender, political orientation, or education on their profiles. Hence, we cannot obtain any demographic information from public profiles of tweeters unless we use a proxy. The prevalence of using swear words is shown to be correlated with gender, social status, and race [12]. Hence, we use LIWC Swear Words category (`swear`) as a measure for demographics feature.

- **Social Engagement:** the more a user is involved with social media the less likely it is for her/him to be misguided by fake news. We measure social engagement on Twitter using the average number of tweets per day.

- **Position in the Network:** this feature can be quantified using a variety of metrics when the network structure is known. However, in the case of social networks between Twitter users, we do not have complete structure and even collecting local information is time consuming due to the rate limitation of Twitter APIs. Hence, we use the information available in our datasets and extract *influence* using the number of followers and *popularity* using the number of followers of each user.

- **Relationship Enhancement:** improving the relation to other social media users and gaining more attention from the community is one of the motivations for spreading fake news. If the number of retweets and likes of a fake news post is higher than the average number of retweets and likes of all the posts by the user, it indicates that this user has enhanced his/her social relations. Hence, we use the difference between the number of retweets and likes of fake news posts and the average values for this user, as the indications of relationship enhancement motivation.

We study the differences between users who spread fake news and the ones who spread real news in terms of five feature categories, as shown in Table 3.1. We set the null hypothesis is that these two groups have the same mean in five categories of features. If the results of T-test shows that there is a significant difference (p-value less than 0.05) between two groups, we can reject the null hypothesis. We observe that: (1) in the Motivational Factors category, we observe a significant difference (p-value < 0.005) between users who share fake news and the ones who share real news in terms of all features except Anxiety; and (2) users also show a significant difference in terms of demographics indicated by the presence of swear words in the tweets. These features can be further utilized to detection fake news by using various classifiers such as random forest, logistic regression, etc.

Table 3.1: Summary of the metrics used to measure psychological user features

Feature Category	Feature Name	Metric	Example Words
Motivational Factors	Tentativeness	LIWC `tentat`	maybe, perhaps
	Discrepancy	LIWC `discrep`	should, would
	Certainty	LIWC `certain`	always, never
	Anxiety	LIWC `anx`	worried, fearful
	Lack of Control	LIWC `focusfuture`	may, will, soon
Demographics	Demographics	LIWC `swear`	damn
Social Engagement	Social Engagement	Avg Tweets per day	-
Position in the Network	Influence	#Folowees	-
	Polularity	#Followers	-
Reationship Enhancement	Boosting #Retweets	Increase in Retweets	-
	Boosting #Likes	Increase in Likes	-

3.1.2   USER BEHAVIOR MODELING

In order to mitigate the effect of fake news, social media sites such as Facebook,[1] Twitter,[2] or Weibo[3] propose allowing users to *flag* a story in their news feed as fake or not, and if the news receives enough flags, it is sent to a trusted third party for manual fact checking. Such flagging behaviors provide useful signals and have the potential to improve fake news detection and reduce the spread of fake news dissemination.

However, the more users are exposed to a story before sending it for fact checking, the greater the confidence a story may be fake news; however, the higher the potential damage if it turns out to be fake news. Thus, it is important to study *when-to-fact-check* problem: finding the optimal timing to collect set of user flags for reducing fake news exposure and improving detection performance. In addition, users may not always flag fake news accurately. Therefore, it is important to consider the reliability of user flagging to decide *what-to-fact-check*: selecting a set of news for fact-checking giving a budget with considering users flagging reliabilities.

When-to-Fact-Check

As in Section 2.4.1, the fact-checking procedure is time consuming and labor intensive, so it is important to optimize the procedure by deciding when to select stories to fact-check [65]. To develop the model that allows the efficient fact-checking procedure, the authors in [65] use the framework of marked temporal point processes and solve a novel stochastic optimal control problem for fact-checking stories in online social networks.

Given an online social networking site with a set of users $\mathcal{U}$ and a set of unverified news posts $\mathcal{C}$ on social media, we define two types of user events: *endogenous events*, which correspond to the publication of stories by users on their own initiative, and *exogenous events*, which correspond to the resharing and/or flagging of stories by users who are exposed to them through their feeds. For a news post, we can characterize the number of **exogenous events** as a counting process $N^e(t)$ with intensity function as $\lambda^e(t)$ which denotes the expected number of exposures at any given time t; and we can use $N^f(t)$ to denote the number of a subset of exposures that involves **flags** by users. Then, we can compute the average number of users exposed to fake news by time t as $\bar{N}(t)$:

$$\bar{N}(t) := p(y = 1|f = 1)N^f(t) + p(y = 1|f = 0)(N^e(t) - N^f(t)), \qquad (3.1)$$

where $p(y = 1|f = 1)$ and $p(y = 1|f = 0)$ denote the conditional probability of a story being fake news (i.e., $y = 1$) given a flag (i.e., $f = 1$) and the conditional probability of a story being fake news given no flag (i.e., $f = 1$). $\bar{N}(t)$ can also be represented as a point process with

---

[1]https://newsroom.fb.com/news/2016/12/news-feed-fyi-addressing-hoaxes-and-fake-news/

[2]https://www.washingtonpost.com/news/the-switch/wp/2017/06/29/twitter-is-looking-for-ways-to-let-users-flag-fake-news

[3]https://www.scmp.com/news/china/policies-politics/article/2055179/how-chinas-highly-censored-wechat-and-weibo-fight-fake

intensity function $\tilde{\lambda}$ as follows:

$$\hat{\lambda}dt = \mathbb{E}_{f(t),f}[d\bar{N}(t)] \tag{3.2}$$

$$= \mathbb{E}_{f(t),f}\Big[(p(y=1|f=1) - p(y=1|f=0))f(t) + p(y=1|f=0)\Big]\lambda^e(t)dt, \tag{3.3}$$

where $f(t)$ checks if there is a flag at time. The probability of flagging a story at time depends on whether the story is fake news or not, which we do not know before we fact-check them. Therefore, we try to estimate this probability using available data and Bayesian statistic. Now we can characterize the problem of fact-checking using above notations as follows:

$$\begin{aligned}
\underset{b(t_0,t_f)}{\text{minimize}} \quad & \mathbb{E}\left[\phi(\hat{\lambda}(t_f)) + \int_{t_0}^{t_f} \ell(\hat{\lambda}(\tau), b(\tau))d\tau\right] \\
\text{subject to} \quad & b(t) \geq 0 \quad \forall t \in (t_0, t_f],
\end{aligned} \tag{3.4}$$

where $b(t)$ is the intensity for the fact-checking scheduling problem that we can optimize, $\phi(\cdot)$ is an arbitrary penalty function, and $\ell(\cdot, \cdot)$ is a loss function which depend on the expected intensity of the spread of fake news $\hat{\lambda}(t_f)$ and the fact-checking intensity, $b(t)$. We can assume the following quadratic forms for $\phi(\cdot)$ and $\ell(\cdot, \cdot)$:

$$\phi(\hat{\lambda}(t_f)) = \frac{1}{2}(\hat{\lambda}(t_f))^2 \tag{3.5}$$

$$\ell(\hat{\lambda}(t), b(t)) = \frac{1}{2}(\hat{\lambda}(t))^2 + \frac{1}{2}qb^2(t), \tag{3.6}$$

where q is a parameter indicating the trade-off between fact-checking and the spread of fake news. Smaller values indicate more resources available for fact-checking and higher values indicate less available resources.

What-to-Fact-Check

Different from the problem of what-to-fact-check in the previous section, the authors in [152] aim to explore, given a set of news, how to select a small set of news pieces for fact-checking to minimize the spreading of fake news. Specifically, they are different in the following aspects: (1) considering the reliability of user flagging behaviors with random variables; (2) agnostic to the actual temporal dynamics of news spreading process; and (3) using discrete epochs with a fixed budget, instead of continuous time with the overall budget. Let $\mathcal{U}$ denote the set of users and $t = 1, 2, \cdots, T$ denote the epochs, where each epoch can be a time window such as one day. The model mainly involves three components: (1) news generation and spreading; (2) users' activity of flagging the news; and (3) selecting news for fact-checking.

At the beginning of each epoch t, the newly coming news set is denoted as $\mathcal{A}^t = \{a_1^t, \cdots, a_N^t\}$. Let the random variable $\mathcal{Y}^t = \{y_1^t, \cdots, y_N^t\}$ denote the set of unknown labels for the news pieces, respectively. Each news a_i^t is associated with a source user who seeded the news, denoted as p_i^t. For each news $a_i^t \in \mathcal{A}^t$, the set of users exposed is initialized as $\pi^{t-1}(s) = \{\}$, and the set of users who flagged the news is initialized as $\psi^{t-1}(a) = \{\}$.

In epoch t, when a news $a \in \mathcal{A}^t$ propagates to a news user $u \in \mathcal{U}$, this user can flag the news to be fake. The set of users who flag news a as fake is denoted by $l^t(a)$. Furthermore, the function $\psi^t(a)$ denotes the complete set of users who have flagged the news a as fake. We can also introduce $\gamma_u \in [0, 1]$ as the probability of user u taking the flagging actions to help annotate fake news, and the bigger γ_u is, the less likely user u will flag fake news. In addition, the accuracy of user labels is conditioned on the news he/she is flagging, represented by two parameters:

- $\alpha_u \in [0, 1]$: the probability that user u flag news a as not fake, given that a is not fake; and

- $\beta_u \in [0, 1]$: the probability that user u flag news a as fake, given that a is fake.

Then we can quantify the observed flagging activity of user u for any news a with the following matrix by variables $(\bar{\theta}_u, \theta_u)$:

$$
\begin{bmatrix} \bar{\theta}_u & 1 - \bar{\theta}_u \\ 1 - \bar{\theta}_u & \theta_u \end{bmatrix} = \gamma \begin{bmatrix} 1 & 1 \\ 0 & 0 \end{bmatrix} + (1 - \gamma_u) \begin{bmatrix} \alpha_u & 1 - \beta_u \\ 1 - \alpha_u 0 & \beta_u \end{bmatrix},
\tag{3.7}
$$

where $\bar{\theta}_u$ and θ_u are defined as follows:

$$
\bar{\theta}_u = p(y_u(a) = 0 | y(a) = 0)
\tag{3.8}
$$

$$
1 - \bar{\theta}_u = p(y_u(a) = 1 | y(a) = 0)
\tag{3.9}
$$

$$
\theta_u = p(y_u(a) = 1 | y(a) = 1)
\tag{3.10}
$$

$$
1 - \theta_u = p(y_u(a) = 0 | y(a) = 1).
\tag{3.11}
$$

At the end of epoch t, we need to select the news set $\mathcal{S}^t$ to send to an expert for acquiring the true labels. If the news is labeled as fake by the expert, the news is blocked to avoid further dissemination. The utility is the number of users *saved* from being exposed to fake news a. The algorithm Ω selects a set $\mathcal{S}^t$ in t, and the total expected utility of the algorithm for $t = 1, \cdots, T$ is given by

$$
\sum_{t=1}^{T} \mathbb{E} \left[\sum_{s \in \mathcal{S}^t} \mathbf{1}_{\{y(s)=1\}} (|\pi^{\infty}(a)| - |\pi^t(a)|) \right],
\tag{3.12}
$$

where $|\pi^{\infty}(a)|$ denotes the number of users who would eventually see the news a, and $|\pi^t(a)|$ means the number of users who have seen news a by the end of epoch t.

To select the optimal set of news $\mathcal{S}^t$ for fact-checking at each epoch t, the proposed algorithm Ω utilizes a Bayesian approach to infer the news labels and learn user parameters through posterior sampling [152].

3.2   POST-BASED DETECTION

Users who are involved in the news dissemination process express their opinions and emotions via posts/comments. These user response provide helpful signals related to the veracity of news claims. Recent research looks into user stance, user emotion, and post credibility to improve the performance of fake news detection. We begin by introducing stance-aggregated modeling.

3.2.1   STANCE-AGGREGATED MODELING

Stances (or viewpoints) indicate the users' opinions toward the news, such as supporting, opposing, etc. Typically, fake news can provoke tremendous controversial views among social media users, in which denying and questioning stances are found to play a crucial role in signaling claims as being fake.

The stance of users' posts can be either explicit or implicit. Explicit stances are direct expressions of emotion or opinion, such as Facebook's "like" actions. Implicit stances can be automatically extracted from social media posts.

Probabilistic Stance Modeling   Consider the scenario where the stances are *explicitly* expressed in "like" actions on social media. Let $A = \{a_1, \cdots, a_j, \cdots, a_N\}$ denote the set of news articles, and $U = \{u_1, \cdots, u_j, \cdots, u_m\}$ represent the set of users engaged in "like" actions. We first construct a bipartite graph $(U \cup A, L)$, where L is the set of likes actions. The idea is that users express "like" actions due to both the user reputations and news qualities. The users and news items can be characterized by the Beta distributions, $\text{Beta}(\alpha_i, \beta_i)$ and $\text{Beta}(\alpha_j, \beta_j)$, respectively. $\text{Beta}(\alpha_i, \beta_i)$ represents the reputation or reliability of user u_i. $\text{Beta}(\alpha_j, \beta_j)$ represents the veracity of news a_j. Intuitively, for a user u_i, $\alpha_i - 1$ represents the number of times u_i likes real news pieces and $\beta_i - 1$ denotes the number of times u_i likes fake news pieces. For a news piece a_j, α_j shows the number of likes a_j receives and β_j means the number of non-likes a_j receives. The expectation values of the Beta distribution are used to estimate the degree of user reputation ($p_i = \frac{\alpha_i}{\alpha_i + \beta_i}$) or news veracity ($p_j = \frac{\alpha_j}{\alpha_j + \beta_j}$). To predict whether a piece of news is fake or not, the linear transformation of p_j is computed: $y_j = 2p_j - 1 = \frac{\alpha_j - \beta_j}{\alpha_j + \beta_j}$, where a positive value indicates true news; otherwise it's fake news.

News Veracity Inference   Let the training set consists of two subsets $A_F, A_T \subseteq A$ for labeled fake and true news, and $\Phi_i = \{u_i | (u_i, a_j) \in L\}$ and $\Phi_j = \{v_j | (u_i, a_j) \in L\}$. The labels are set as $y_j = -1$ for all $a_j \in A_F$, and $y_j = 1$ for all $a_j \in A_T$, and $y_j = 0$ for unlabeled news pieces. The parameter optimization of user u_i is performed iteratively by following updating functions:

$$\alpha_i = \Delta_\alpha + \sum_{y_i>0, i\in\Phi_i} y_i$$

$$\beta_i = \Delta_\beta - \sum_{y_i<0, i\in\Phi_i} y_i \qquad (3.13)$$

$$y_i = (\alpha_i - \beta_i)/(\alpha_i + \beta_i),$$

where Δ_α and Δ_β are the prior base constants indicating the degree the user believing the fake or true news. Similarly, the parameter of news a_j is updated as

$$\alpha_j = \Delta'_\alpha + \sum_{y_j>0, j\in\Phi_j} y_j$$

$$\beta_j = \Delta'_\beta - \sum_{y_j<0, j\in\Phi_j} y_j \qquad (3.14)$$

$$y_j = (\alpha_j - \beta_j)/(\alpha_j + \beta_j),$$

where Δ'_α and Δ'_β are the prior constants indicating the ratio of fake or true news. In this way, the stance (like) information is aggregated to optimize the parameters, which can be used to predict the news veracity using y_j [147].

We can infer the *implicit* stance values from social media posts, which usually requires a labeled stance dataset to train a supervised model. The inferred stance scores then serve as the input to perform fake news classification.

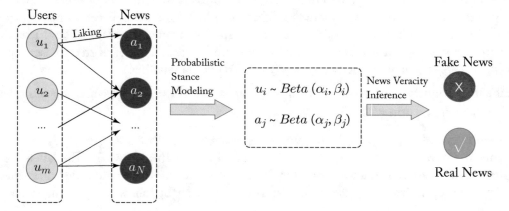

Figure 3.3: The illustration of stance aggregation framework: (1) probabilistic stance modeling and (2) news veracity inference.

3.2.2 EMOTION-ENHANCED MODELING

Fake news publishers often aims to spread information extensively and draw wide public attention. Long-standing social science studies demonstrate that the news which evokes high-arousal, or activating (awe, anger, or anxiety) emotions is more viral on social media [42, 146]. To achieve this goal, fake news publishers commonly adopt two approaches. First, publishers post news with intense emotions which trigger a high level of physiological arousal in the crowd. For example, in Figure 3.4a, the publisher uses rich emotional expressions (e.g., `Oh my god!`) to make this information more impressive and striking. Second, publishers may present the news objectively to make it convincing whose content, however, is controversial which evoke intense emotion in the public, and finally spreads widely. As another example (see Figure 3.4b), the publisher writes the post in a unemotional way, while the statement that `China ranks second to the last` suddenly bring on tension in the crowd, and people express their feeling of anger (e.g., `most ridiculous`), shock, and doubt (e.g., `seriously?`) in comments.

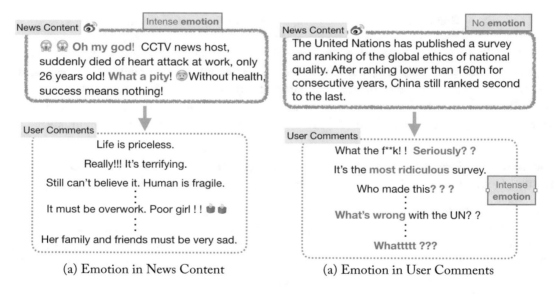

(a) Emotion in News Content (a) Emotion in User Comments

Figure 3.4: Two fake news posts from Sina Weibo. (a) A post which contains emotions of astonishment and sadness in **news content** that easily arouses the audience. (b) A post which contains no emotion, but raises emotions like doubt and anger in **user comments** by controversial topics. Based on [47].

The end-to-end emotion based fake news detection framework (see Figure 3.5) consists of three major components: (i) the content module mine the information from the publisher, including semantics and emotions in news contents; (ii) the comment module capture semantics

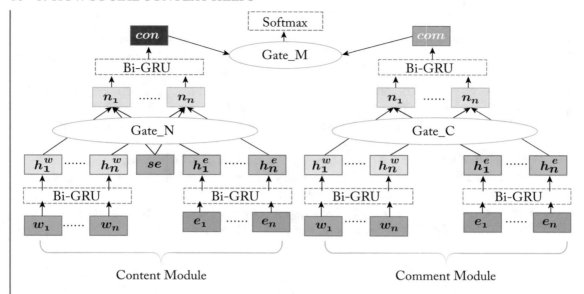

Figure 3.5: The proposed framework consists of three components: (1) the news content module; (2) the user comments module; and (3) the fake news prediction component. The previous two modules are used to model semantics and emotions from the publisher and users, respectively, while the prediction part fuses information of these two modules and makes prediction. Three gates (Gate_N, Gate_C, and Gate_M) are used for multi-modal fusion in different layers.

and emotion information from users; and (iii) the fake news prediction component fuses the features from both news content and user comments and predict fake news.

Learning Emotion Embeddings   Early studies primarily use hand-crafted features for representing emotion of text, which highly rely on sentiment dictionaries. There are several widely used emotion dictionaries such as WordNet [62], SlangSD [170], and MPQA [168] for English and HowNet[4] for Chinese. However, this method may encounter problems of emotion migration and low coverage on social media, because of the differences of sentiment word usage on social media and in the real word. In addition, some existing tools such as Vader [55] are designed to predict sentiment for a general purpose on social media, which may not be specific for fake news detection and the resultant numeric sentiment score is not easily embedded to deep learning models.

Therefore, we adopt the deep learning emotion prediction model [4] to learn the task-specific sentiment embedding for both news contents and user comments. Inspired by recent advancements on deep learning for emotion modeling [4], we train a recurrent neural network

[4]http://www.keenage.com/html/e_index.html

(RNN) to learn the emotion embedding vectors. Following traditional settings [54], we first obtain a large-scale real-world datasets that contain emotions, and use the emotions as the emotion labels, and then initialize each word with one-hot vector. After initiation, all word vectors pass an embedding layer which project each words from the original one-hot space into a low dimensional space, and then sequentially fed into a one-layer GRU model. Then, through back-propagation, the embedding layer get updated during training, producing emotion embedding $\mathbf{e}_i$ for each word w_i.

Incorporating Emotion Representations   We introduce how to incorporate emotion embeddings to news contents and user comments to learn the representations for fake news detection. We can learn the basic textual feature representations through a bidirectional GRU word encoder as in Section 2.1.3. For each word w_i, the word embedding vector $\mathbf{w}_i$ is initialized with the pre-trained word2vec [90]. The bidirectional GRU contains the forward GRU $\overrightarrow{f}$ which reads each sentence from word w_0 to w_M and a backward GRU $\overleftarrow{f}$ which reads the sentence from word w_n to w_0:

$$\overrightarrow{\mathbf{h}_i^w} = \overrightarrow{GRU}(\mathbf{w}_i), i \in [0, n],$$
$$\overleftarrow{\mathbf{h}_i^w} = \overleftarrow{GRU}(\mathbf{w}_i), i \in [0, n]. \tag{3.15}$$

For a given word w_i, we could obtain its word encoding vector $\mathbf{h}_i^w$ by concatenating the forward hidden state $\overrightarrow{\mathbf{h}_i^w}$ and backward hidden state $\overleftarrow{\mathbf{h}_i^w}$, i.e., $\mathbf{h}_i^w = [\overrightarrow{\mathbf{h}_i^w}, \overleftarrow{\mathbf{h}_i^w}]$.

Similarly to the word encoder, we adopt bidirectional GRU to model the emotion feature representations for the words. After we obtain the emotion embedding vectors $\mathbf{e}_i$, we can learn the emotion encoding $\mathbf{h}_i^e$ for word w_i:

$$\overrightarrow{\mathbf{h}_i^e} = \overrightarrow{GRU}(\mathbf{e}_i), i \in [0, n],$$
$$\overleftarrow{\mathbf{h}_i^e} = \overleftarrow{GRU}(\mathbf{e}_i), i \in [0, n], \tag{3.16}$$

for a given word w_i, we could obtain its emotion encoding vector $\mathbf{h}_i^e$ by concatenating the forward hidden state $\overrightarrow{\mathbf{h}_i^e}$ and backward hidden state $\overleftarrow{\mathbf{h}_i^e}$, i.e., $\mathbf{h}_i^e = [\overrightarrow{\mathbf{h}_i^e}, \overleftarrow{\mathbf{h}_i^e}]$.

The overall emotion information of news content is also important when deciding how much information from emotion embedding should be absorbed for the words. For a given post a, we extract the emotion features included in [22] and also add some emotion features. There are 19 features regarding emotion aspects of news, including *numbers of positive/negative words, sentiment score*, etc. News emotion features of a is denoted as se.

Gate_N is applied to learn information jointly from word embedding, emotion embedding and sentence emotion features, and yield new representation for each word (see Figure 3.5). The units in Gate_N is motivated by the *forget gate* and *input gate* in LSTM. In Gate_N, two emotion inputs corporately decide the value of r_t and u_t with two sigmoid layers, which are used for manage how much information from semantic and emotion is added into the new representation. Meanwhile, a dense layer transfer the emotion inputs to the same dimensional space of word embedding. Mathematically, the relationship between inputs and output of T_Gate is defined as the following formulas:

$$\begin{aligned}
\mathbf{r}_t &= \sigma(\mathbf{W}_r \cdot [se, \mathbf{h}_t^e] + \mathbf{b}_r) \\
\mathbf{u}_t &= \sigma(\mathbf{W}_u \cdot [se, \mathbf{h}_t^e] + \mathbf{b}_u) \\
\mathbf{c}_t^e &= \tanh(\mathbf{W}_c \cdot [se, \mathbf{h}_t^e] + \mathbf{b}_c) \\
\mathbf{n}_t &= \mathbf{r}_t \odot \mathbf{h}_t^w + \mathbf{u}_t \odot \mathbf{c}_t^e.
\end{aligned} \tag{3.17}$$

Comment module explore the semantic and emotion information from the users in the event. The architecture of comment module is similar to content module's except: (1) all comments are first concatenated before fed into BiGRUs; (2) there is no sentence emotion features; and (3) Gate_C is used for fusion. Gate_C is introduced for fusion in comment module. Different from Gate_N, there are only two modalities. We adopt the *update gate* in GRU to control the update of information in fusion process (see Figure 3.5). Two inputs jointly yield a update gate vector u_t through a sigmoid layer. A dense layer create a vector of new candidate values, h_t^e, which has the same dimension as the w_t. The final output n_t is a linear interpolation between the w_t and h_t^e. Mathematically, the following formulas represent the process:

$$\begin{aligned}
\mathbf{u}_t &= \sigma(\mathbf{W}_u \cdot [\mathbf{h}_t^w, \mathbf{h}_t^e] + \mathbf{b}_u) \\
\mathbf{c}_t^e &= \tanh(\mathbf{W}_c \cdot \mathbf{h}_t^e + \mathbf{b}_c) \\
\mathbf{n}_t &= \mathbf{u}_t \odot \mathbf{h}_t^w + (1 - \mathbf{u}_t) \odot \mathbf{c}_t^e.
\end{aligned} \tag{3.18}$$

Emotion-Based Fake News Detection   Here, Gate_M fuse the high-level representation of content module and comment module, and then yield a representation vector n (see Figure 3.5). Mathematically, following equations demonstrate the internal relationship of Gate_M:

$$\begin{aligned}
\mathbf{r} &= \sigma(\mathbf{W}_u \cdot [con, com] + \mathbf{b}_u) \\
\mathbf{o} &= \mathbf{r} \odot con + (1 - \mathbf{r}) \odot com.
\end{aligned} \tag{3.19}$$

We use a fully connected layer with softmax activation to project the new vector n into the target space of two classes: fake news and real news, and gain the probability distribution:

$$\hat{\mathbf{y}} = \mathrm{softmax}(\mathbf{W}_f \mathbf{o} + \mathbf{b}_f), \tag{3.20}$$

where $\hat{\mathbf{y}} = [\hat{\mathbf{y}}_0, \hat{\mathbf{y}}_1]$ is the predicted probability vector with $\hat{\mathbf{y}}_0$ and $\hat{\mathbf{y}}_1$ indicate the predicted probability of label being 0 (real news) and 1 (fake news), respectively. $y \in \{0, 1\}$ denotes the ground truth label of news. $\mathbf{b}_f \in \mathbb{R}^{1 \times 2}$ is the bias term. Thus, for each news piece, the goal is to minimize the cross-entropy loss function as follows:

$$\mathcal{L}(\theta) = -y \log(\hat{\mathbf{y}}_1) - (1 - y) \log(1 - \hat{\mathbf{y}}_0), \tag{3.21}$$

where θ denotes the parameters of the network.

3.2.3 CREDIBILITY-PROPAGATED MODELING

Credibility-propagated models aims to infer the veracity of news pieces from the credibility of the posts on social media through network propagation. The basic assumption is that the credibility of a given news event is highly related to the credibility degree of its relevant social media posts [59]. Since posts are correlated in terms of their viewpoints toward the news piece, we need to collect all relevant social media posts and represent a credibility network among them. Then, we can explore the correlations among posts to optimize the credibility values, which can be averaged as the score for predicting fake news.

Representing a Credibility Network

We can first build a credibility network structure among all the posts $C = \{c_1, \cdots, c_m\}$ for a news piece a (see Figure 3.6). Credibility network initialization consists of two parts: node initialization and link initialization. First, we can obtain the initial credibility score vector of nodes $\mathbf{T_0}$ from pre-trained classifiers with features extracted from external training data. The link is defined by mining the viewpoint relations, which are the relations between each pair of viewpoint such as contradicting or same. The basic idea is that posts with same viewpoints form supporting relations which raise their credibilities, and posts with contradicting viewpoints form opposing relations which weaken their credibilities. Specifically, a social media post c_i is modeled as a multinomial distribution θ_i over K topics, and a topic k is modeled as a multinomial distribution ψ_{tk} over L viewpoints. The probability of a post c_t over topic k along with L viewpoints is denoted as $p_{ik} = \theta_i \times \psi_{ik}$. The distance between two posts c_i and c_j are measured by using the Jensen–Shannon Distance: $Dis(c_i, c_j) = D_{JS}(p_{ik}||p_{jk})$.

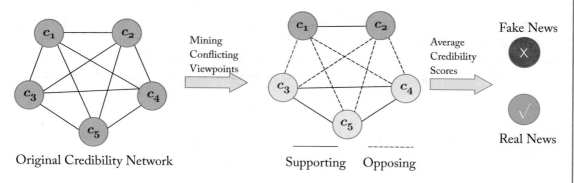

Figure 3.6: An illustration of leveraging post credibility to detect fake news.

The supporting or opposing relation indicator is determined as follows: it's assumed that one post contains a major topic-viewpoint, which can be defined as the largest proportion of p_{ik}. If the major topic-viewpoints of two posts c_i and c_j are clustered together (they take the same viewpoint), then they are mutually supporting; otherwise, they are mutually opposing. The similarity/dissimilarity measure of two posts are defined as:

$$f(c_i, c_j) = \frac{(-1)^b}{D_{JS}(p_{ik}||p_{jk}) + 1},$$

(3.22)

where b is the link type indicator, and if $b = 0$, then c_i and c_j take the same viewpoint; otherwise, $b = 1$.

Propagating Credibility Values

The goal is to optimize the credibility values of each node (i.e., social media post), and infer the credibility value of corresponding news items [59]. Posts with supporting relations should have similar credibility values; posts with opposing relations should have opposing credibility values. In the credibility network, there are: (i) a post credibility vector $\mathbf{T} = \{o(c_1), o(c_2), ..., o(c_n)\}$ with $o(c_i)$ denoting the credibility value of post c_i; and (ii) a matrix $\mathbf{W} \in \mathbb{R}^{n \times n}$, where $\mathbf{W}_{ij} = f(c_i, c_j)$ which denotes the *viewpoint* correlations between post c_i and c_j, that is, whether the two posts take supporting or opposing positions. Therefore, the objective to propagate credibility scores can be defined as a network optimization problem as below:

$$Q(\mathbf{T}) = \mu \sum_{i,j=1}^{n} |\mathbf{W}_{ij}| \left(\frac{o(c_i)}{\sqrt{\bar{\mathbf{D}}_{ii}}} - e_{ij} \frac{o(c_j)}{\sqrt{\bar{\mathbf{D}}_{jj}}} \right)^2$$
$$+ (1 - \mu)\|\mathbf{T} - \mathbf{T}_0\|^2,$$

(3.23)

where $\bar{\mathbf{D}}$ is a diagonal matrix with $\bar{\mathbf{D}}_{ii} = \sum_k |\mathbf{W}_{ik}|$ and $e_{ij} = 1$, if $\mathbf{W}_{ij} \geq 0$; otherwise $e_{ij} = 0$. The first component is the smoothness constraint which guarantees the two assumptions of supporting and opposing relations; the second component is the fitting constraint to ensure variables not change too much from their initial values; and μ is the regularization parameter. Then the credibility propagation on the proposed network G_C is formulated as the minimization of this loss function:

$$\mathbf{T}^* = \text{argmin}_{\mathbf{T}} Q(\mathbf{T}).$$

(3.24)

The optimum solution can be solved by updating $\mathbf{T}$ in an iterative manner through the transition function $\mathbf{T}(t) = \mu \mathbf{H} \mathbf{T}(t-1) + (1 - \mu \mathbf{T}_0)$, where $\mathbf{H} = \bar{\mathbf{D}}^{-1/2} \mathbf{W} \bar{\mathbf{D}}^{-1/2}$. As the iteration converges, each post receives a final credibility value, and the average of them is served as the final credibility evaluation result for the news.

3.3   NETWORK-BASED DETECTION

Recent advancements of network representation learning, such as network embedding and deep neural networks, allow us to better capture the features of news from auxiliary information such as friendship networks, temporal user engagements, and interaction networks. Network-based fake news detection aims to leverage the advanced network analysis and modeling methods to better predict fake news. We introduce representative types of networks for detecting fake news.

3.3.1   REPRESENTATIVE NETWORK TYPES

We introduce several network structures that are commonly used to detect fake news (Figure 3.7).

Friendship Networks   A user's friendship network is represented as a graph $G_F = (\mathcal{U}, E_F)$, where $\mathcal{U}$ and E_F are the node and edge sets, respectively. A node $u \in \mathcal{U}$ represents a user, and $(u_1, u_2) \in E$ represents whether a social relation exists.

Homophily theory [87] suggests that users tend to form relationships with like-minded friends, rather than with users who have opposing preferences and interests. Likewise, social influence theory [85] predicts that users are more likely to share similar latent interests toward news pieces. Thus, the friendship network provides the structure to understand the set of social relationships among users. The friendship network is the basic route for news spreading and can reveal community information.

Diffusion Networks   A diffusion network is represented as a directed graph $G_D = (\mathcal{U}, E_D, p, t)$, where $\mathcal{U}$ and E are the node and edge sets, respectively. A node $u \in \mathcal{U}$ represents an individual that can publish, receive, and diffuse information at time $t_i \in t$. A directed edge, $(u_1 \rightarrow u_2) \in E_D$, between nodes $u_1, u_2 \in \mathcal{U}$, represents the direction of information diffusion. Each directed edge $(u_1 \rightarrow u_2) \in E_D$, between nodes $u_1, u_2 \in U$, represents the direction of information diffusion. Each directed edge $(u_1 \rightarrow u_2)$ is assumed to be associated with an information diffusion probability, $p(u_1 \rightarrow u_2) \in [0, 1]$.

The diffusion network is important for learning about representations of the structure and temporal patterns that help identify fake news. By discovering the sources of fake news and the spreading paths among the users, we can also try to mitigate the fake news problem.

Interaction Networks   An interaction network $G_I = (\{\mathcal{P}, \mathcal{U}, \mathbf{A}\}, E_I)$ consists of nodes representing publishers, users, news, and the edges E_I indicating the interactions among them. For example, edge $(p \rightarrow a)$ demonstrates that publisher p publishes news item a, and $(v \rightarrow u)$ represents news a is spread by user u.

The interaction networks can represent the correlations among different types of entities, such as publisher, news, and social media post, during the news dissemination process [141]. The characteristics of publishers and users, and the publisher-news and news-users interactions have potential to help differentiate fake news.

Propagation Networks   A propagation network $G_P = (\mathcal{C}, a)$ consists of a news piece a and the corresponding social media posts $\mathcal{C}$ that propagates the news. Note that different types of posts can occurs such as reposting, replying, commenting, liking, etc. We will introduce that a

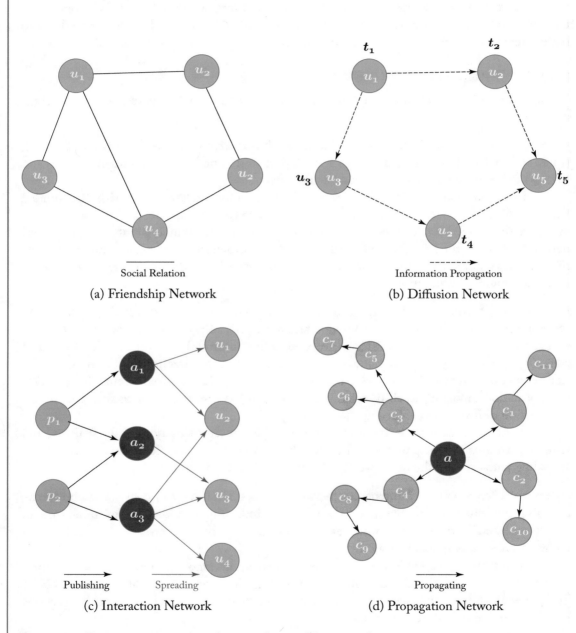

Figure 3.7: Representative network types during fake news dissemination.

propagation network is treated in a hierarchical view, consisting of two levels: *macro-level* and *micro-level*. The macro-level propagation network includes the news nodes, tweet nodes, and retweet nodes. The micro-level propagation network indicates the conversation tree represented by reply nodes.

Propagation networks contain rich information from different perspectives such as temporal, linguistic, and structural, which provides auxiliary information for potential improving the detecting of fake news.

3.3.2   FRIENDSHIP NETWORKING MODELING

The friendship network plays an important role in fake news diffusion. The fact that users are likely to form echo chambers strengthens our need to model user social representations and to explore its added value for a fake news study. Essentially, given the friendship network G_F, we want to learn latent representations of users while preserving the structural properties of the network, including first-order and higher-order structure, such as second-order structure and community structure. For example, Deepwalk [108] preserves the neighborhood structure of nodes by modeling a stream of random walks. In addition, LINE [150] preserves both first-order and second-order proximities. Specifically, we measure the first-order proximity by the joint probability distribution between the user u_i and u_j,

$$p_1(u_i, u_j) = \frac{1}{1 + \exp(-\mathbf{u_i}^T \mathbf{u_j})}, \qquad (3.25)$$

where u_i (u_j) is the social representation of user u_i (u_j). We model the second-order proximity by the probability of the context user u_j being generated by the user u_i, as follows:

$$p_2(u_j | u_i) = \frac{\exp(\mathbf{u_j}^T \mathbf{u_i})}{\sum_{k=1}^{|V|} \exp(\mathbf{u_k}^T \mathbf{u_i})}, \qquad (3.26)$$

where $|V|$ is the number of nodes or "contexts" for user u_i. This conditional distribution implies that users with similar distributions over the contexts are similar to each other. The learning objective is to minimize the KL-divergence of the two distributions and empirical distributions, respectively.

Network communities may actually be a more important structural dimension because fake news spreaders are likely to form polarized groups [49, 136]. This requires the representation learning methods to be able to model community structures. For example, a community-preserving node representation learning method, Modularized Nonnegative Matrix Factorization (MNMF), is proposed [164]. The overall objective is defined as follows:

$$\min_{\mathbf{M},\mathbf{U},\mathbf{H},\mathbf{C} \geq 0} \underbrace{\| \mathbf{S} - \mathbf{M}\mathbf{U}^T \|_F^2}_{\text{Proximity Mapping}} + \underbrace{\alpha \| \mathbf{H} - \mathbf{U}\mathbf{C}^T \|_F^2}_{\text{Community Mapping}} - \underbrace{\beta tr(\mathbf{H}^T \mathbf{B}\mathbf{H})}_{\text{Modularity Modeling}} \qquad (3.27)$$
$$\text{s.t. } tr(\mathbf{H}^T \mathbf{H}) = m$$

and comprises three major parts: proximity mapping, community mapping, and modularity modeling. In proximity mapping, $\mathbf{S} \in \mathbb{R}^{m \times m}$ is the user similarity matrix constructed from the user adjacency matrix (first-order proximity) and neighborhood similarity matrix (second-order proximity), and $\mathbf{M} \in \mathbb{R}^{m \times k}$ and $\mathbf{U} \in \mathbb{R}^{m \times k}$ are the basis matrix and user representations. For community mapping, $\mathbf{H} \in \mathbf{R}^{m \times l}$ is the user-community indicator matrix that we optimize to be reconstructed by the product of the user latent matrix $\mathbf{U}$ and the community latent matrix $\mathbf{C} \in \mathbf{R}^{l \times m}$. For modularity modeling, the objective is to maximize the modularity function [96], where $\mathbf{B} \in \mathbb{R}^{m \times m}$ is the modularity matrix.

Tensor factorization can be applied to learn the community-enhanced news representation to predict fake news [49]. The goal is to incorporate user community information to guide the learning process of news representation. We first build a three-mode news-user-community tensor $\underline{\mathbf{Y}} \in \mathbb{R}^{N \times m \times J}$. Then we apply the CP/PARAFAC tensor factorization model to factorize $\underline{\mathbf{Y}}$ into the following:

$$\underline{\mathbf{Y}} \approx [\mathbf{F}, \mathbf{U}, \mathbf{H}] = \sum_{r=1}^{R} \lambda_r \mathbf{f}_r \odot \mathbf{u}_r \odot \mathbf{h}_r, \tag{3.28}$$

where $\odot$ denotes the outer product and $\mathbf{f}_r$ (same for $\mathbf{b}_r$ and $\mathbf{h}_r$) denotes the normalized rth column of non-negative factor matrix $\mathbf{F}$ (same for $\mathbf{B}$ and $\mathbf{H}$), and R is the rank. Each row of $\mathbf{F}$ denotes the representation of the corresponding article in the embedding space.

3.3.3 DIFFUSION NETWORK TEMPORAL MODELING

The news diffusion process involves abundant temporal user engagements on social media [119, 136, 169]. The social news engagements are defined as a set of tuples to represent the process of how news items spread over time among m users in $\mathcal{U} = \{u_1, u_2, ..., u_m\}$. Each engagement $\{u_i, t_i, c_i\}$ represents that a user u_i spreads news article at time t_i by posting c_i. For example, a diffusion path between two users u_i and u_j exists if and only if: (1) u_j follows u_i and (2) u_j posts about a given news only after u_i does so.

The goal of learning temporal representations is to capture the user's pattern of temporal engagements with a news article a_j. Recent advances in the study of deep neural networks, such as RNNs, have shown promising performance for learning representations. RNNs are powerful structures that allow the use of loops within the neural network to model sequential data. Given the diffusion network G_D, the key procedure is to construct meaningful features $\mathbf{x}_i$ for each engagement. The features are generally extracted from the contents of c_i and the attributes of u_i. For example, $\mathbf{x}_i$ consists of the following components: $\mathbf{x}_i = (\eta, \Delta t, \mathbf{u}_i, \mathbf{c}_i)$. The first two variables η and Δt represent the number of total user engagements through time t and the time difference between engagements, respectively. These variables capture the general measure of frequency and time interval distribution of user engagements of the news piece a_j. For the content features of users posts, the $\mathbf{c}_i$ are extracted from hand-crafted linguistic features, such as n-gram features, or by using word embedding methods such as doc2vec [76] or GloVe [106].

We extract the features of users $\mathbf{u}_i$ by performing a singular value decomposition of the user-news interaction matrix $\mathbf{E} \in \{0, 1\}^{m \times N}$, where $\mathbf{E}_{ij} = 1$ indicate that user u_i has engaged in the process of spreading the news piece a_j; otherwise $\mathbf{E}_{ij} = 0$.

An RNN framework for learning news temporal representations is demonstrated in Figure 3.8. Since $\mathbf{x}_i$ includes features that come from different information space, such as temporal and content features, so we do not suggest incorporating $\mathbf{x}_i$ into RNN as the raw input. Thus, we add a fully connected embedding layer to convert the raw input $\mathbf{x}_i$ into a standardized input features $\tilde{\mathbf{x}}_i$, in which the parameters are shared among all raw input features $\mathbf{x}_i, i = 1, ..., m$. Thus, the RNN takes a sequence $\tilde{\mathbf{x}}_1, \tilde{\mathbf{x}}_2, ..., \tilde{\mathbf{x}}_m$ as input. At each time-step i, the output of previous step $\mathbf{h}_{i-1}$, and the next feature input $\tilde{\mathbf{x}}_i$ are used to update the hidden state $\mathbf{h}_i$. The hidden states $\mathbf{h}_i$ is the feature representation of the sequence up to time i for the input engagement sequence. Thus, the hidden states of final step $\mathbf{h}_m$ is passed through a fully connected layer to learn the resultant news representation, defined as $\mathbf{a}_j = \tanh(\mathbf{W}\mathbf{h}_m + \mathbf{b})$, $\mathbf{b}$ is a bias vector. Thus, we can use $\mathbf{a}_j$ to perform fake news detection and related tasks [119].

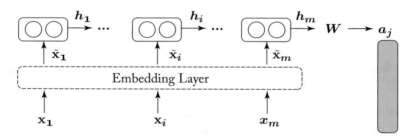

Figure 3.8: An RNN framework for learning news temporal representations.

3.3.4 INTERACTION NETWORK MODELING

Interaction networks describe the relationships among different entities such as publishers, news pieces, and users. Given the interaction networks the goal is to embed the different types of entities into the same latent space, by modeling the interactions among them. We can leverage the resultant feature representations of news to perform fake news detection. The framework is shown as in Figure 3.9, which mainly includes the following components: a news contents embedding component, a user embedding component, a user-news interaction embedding component, a publisher-news relation embedding component, and a semi-supervised classification component. In general, the news contents embedding component describes the mapping of news from bag-of-word features to latent feature space; the user embedding component illustrates the extraction of user latent features from user social relations; the user-news interaction embedding component learn the feature representations of news pieces guided by their partial labels and user credibilities. The publisher-news relation embedding component regularizes the fea-

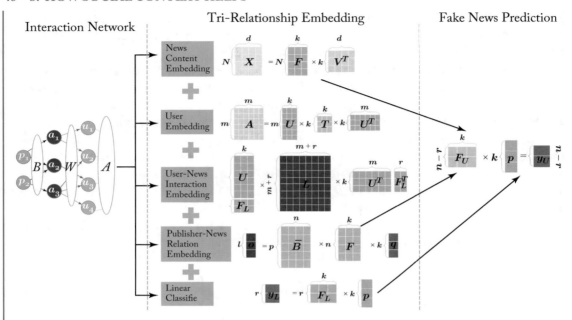

Figure 3.9: The framework for interactive network embedding for fake news detection.

ture representations of news pieces through publisher partisan bias labels. The semi-supervised classification component learns a classification function to predict unlabeled news items.

News Embedding   We can use news content to find clues to differentiate fake news and true news. As in Equation (3.27) from Section 2.4.1, we use NMF we can attempt to project the document-word matrix to a joint latent semantic factor space with low dimensionality, such that the document-word relations are modeled as the inner product in the space. We can obtain the news representation matrix $\mathbf{F}$.

User Embedding   On social media, people tend to form relationships with like-minded friends, rather than with users who have opposing preferences and interests [140]. Thus, connected users are more likely to share similar latent interests in news pieces. To obtain a standardized representation, we use nonnegative matrix factorization to learn the user's latent representations (we will introduce other methods in Section 3.3.2). Specifically, giving user-user adjacency matrix $\mathbf{A} \in \{0, 1\}^{m \times m}$, we learn nonnegative matrix $\mathbf{U} \in \mathbb{R}_+^{m \times k}$ by solving the following optimization problem:

$$\min_{\mathbf{U}, \mathbf{T} \geq 0} \|\mathbf{Y} \odot (\mathbf{A} - \mathbf{U}\mathbf{T}\mathbf{U}^T)\|_F^2, \tag{3.29}$$

where $\mathbf{U}$ is the user latent matrix, $\mathbf{T} \in \mathbb{R}_+^{k \times k}$ is the user-user correlation matrix, and $\mathbf{Y} \in \mathbb{R}^{m \times m}$ controls the contribution of $\mathbf{A}$. Since only positive samples are given in $\mathbf{A}$, we can first set $\mathbf{Y} =$

$sign(\mathbf{A})$, then perform negative sampling and generate the same number of unobserved links and set weights as 0.

User-News Embedding   The user-news interactions are often modeled by considering the relationships between user representations and the news veracity values ($\mathbf{y}_{Lj}$). Intuitively, users with low credibilities are more likely to spread fake news, while users with high credibility scores are less likely to spread fake news. Each user has a credibility score that we can infer using his/her published posts [1], and we use $\mathbf{s} = \{\mathbf{s}_1, \mathbf{s}_2, ..., \mathbf{s}_m\}$ to denote the credibility score vector, where a larger $\mathbf{s}_i \in [0, 1]$ indicates that user u_i has a higher credibility. The user-news engaging matrix is represented as $\mathbf{E} \in \{0, 1\}^{m \times N}$, where $\mathbf{E}_{ij} = 1$ indicates that user u_i has engaged in the spreading process of the news piece a_j; otherwise $\mathbf{E}_{ij} = 0$. The objective function is shown as follows:

$$
\min \ \underbrace{\sum_{i=1}^{m} \sum_{j=1}^{r} \mathbf{E}_{ij} \mathbf{s}_i \left(1 - \frac{1 + \mathbf{y}_{Lj}}{2} \right) ||\mathbf{U}_i - \mathbf{F}_j||_2^2}_{\text{True news}}
$$

$$
+ \underbrace{\sum_{i=1}^{m} \sum_{j=1}^{r} \mathbf{E}_{ij} (1 - \mathbf{s}_i) \left(\frac{1 + \mathbf{y}_{Lj}}{2} \right) ||\mathbf{U}_i - \mathbf{F}_j||_2^2}_{\text{Fake news}},
$$

(3.30)

where $\mathbf{y}_L \in \mathbb{R}^{r \times 1}$ is the label vector of all partially labeled news. The objective considers two situations: (i) for true news, i.e., $\mathbf{y}_{Lj} = -1$, which ensures that the distance between latent features of high-credibility users and that of true news is small; and (ii) for fake news, i.e., $\mathbf{y}_{Lj} = 1$, which ensures that the distance between the latent features of low-credibility users and the latent representations of true news is small.

Publisher-News Embedding   The publisher-news interactions are modeled by incorporating the characteristics of the publisher and news veracity values (). Fake news is often written to convey opinions or claims that support the partisan bias of the news publisher. Publishers with a high degree of political bias are more likely to publish fake news [141]. Thus, a useful news representation should be good for predicting the partisan bias score of its publisher. The partisan bias scores are collected from fact-checking websites and are represented as a vector $\mathbf{o}$. We utilize publisher partisan labels vector $\mathbf{o} \in \mathbb{R}^{l \times 1}$ and publisher-news matrix $\mathbf{B} \in \mathbb{R}^{l \times N}$ to optimize the news feature representation learning as follows:

$$
\min \ || \bar{\mathbf{B}} \mathbf{F} \mathbf{Q} - \mathbf{o} ||_2^2,
$$

(3.31)

where the latent features of a news publisher are represented by the features of all the news he/she published, i.e., $\bar{\mathbf{B}} \mathbf{D}$. $\bar{\mathbf{B}}$ is the normalized user-news publishing relation matrix, i.e., $\bar{\mathbf{B}}_{kj} = \frac{\mathbf{B}_{kj}}{\sum_{j=1}^{n} \mathbf{B}_{kj}}$. $\mathbf{Q} \in \mathbb{R}^{k \times 1}$ is the weighting matrix that maps news publishers' latent features to corresponding partisan label vector $\mathbf{o}$.

The finalized model combines all previous components into a coherent model. In this way, we can obtain the latent representations of news items $\mathbf{F}$ and of users $\mathbf{U}$ through the network embedding procedure, which is utilized to perform fake news classification tasks.

3.3.5   PROPAGATION NETWORK DEEP-GEOMETRIC MODELING

In [34, 92], the authors propose to use geometric deep learning (e.g., graph convolution neural networks) to learn the structural of propagation networks for fake news detection. Geometric deep learning naturally deals with heterogeneous graph data, which has the potential to unify signals of text and structure information in the propagation networks. Geometric deep learning generally refers to the non-Euclidean deep learning approaches [20]. In general, graph CNNs replace the classical convolution operation on grids with a local permutation-invariant aggregation on the neighborhood of a vertex in a graph. Specifically, the convolution works with a spectral representation of the graphs G_P and learns the spatially localized filters by approximating convolutions defined on the graph Fourier domain. Mathematically, a normalized graph laplacian $\mathbf{L}$ is defined as $\mathbf{L} = \mathbf{I}_N - \mathbf{D}^{-\frac{1}{2}}\mathbf{A}\mathbf{D}^{-\frac{1}{2}} = \mathbf{E}\Lambda\mathbf{E}^T$, where $\mathbf{D}$ is the degree matrix of the adjacency matrix $\mathbf{A}$ for the propagation network ($\mathbf{D}_{ii} = \sum_j \mathbf{A}_{ij}$). Λ is the diagonal matrix of its eigenvalues and $\mathbf{E}$ is the matrix of eigenvector basis. Given a node feature $\mathbf{c}$, $\mathbf{E}^T\mathbf{c}$ is the graph Fourier transform of x. The convolutional operation on this node signal is defined as:

$$g_\theta * \mathbf{c} = \mathbf{E}g_\theta\mathbf{E}^T\mathbf{c}, \tag{3.32}$$

where $g_\theta = diag(\theta)$ parameterized by θ is a function of the eigenvalues of $\mathbf{L}$, i.e., $g_\theta(\Lambda)$. However, convolution in Equation (3.32) is computationally expensive due to the multiplication with high dimensional matrix $\mathbf{E}$ and it is a non-spatially localized filters. To solve this problem, it is suggested to use Chebyshev polynomials $T_k(\mathbf{c})$ up to K^{th} order as a truncated expansion to approximate g_θ. Equation (3.32) thus is reformulated as:

$$g_\theta * \mathbf{c} \approx \sum_{k=0}^{K} \theta_k T_k(\tilde{\mathbf{L}})\mathbf{c}. \tag{3.33}$$

$\tilde{\mathbf{L}} = \frac{2}{\lambda_{max}}\mathbf{L} - \mathbf{I}_N$ and λ_{max} is the largest eigenvalue of $\mathbf{L}$. Now, θ_k becomes the Chebyshev coefficients. If we limit $K = 1$ and approximate $\lambda_{max} \approx 2$, with the normalized tricks and weak constraints used in [68], Equation (3.33) simplifies to:

$$g_\theta * \mathbf{c} \approx \theta \left(\tilde{\mathbf{D}}^{-\frac{1}{2}}\tilde{\mathbf{A}}\tilde{\mathbf{D}}^{-\frac{1}{2}}\right)\mathbf{c}, \tag{3.34}$$

where $\tilde{\mathbf{A}} = \mathbf{A} + \mathbf{I}_N$ and $\tilde{\mathbf{D}}$ is the degree matrix of $\tilde{\mathbf{A}}$. We turn Equation (3.34) to the matrix multiplication form, for the whole network,

$$\mathbf{C}' = \delta \left(\tilde{\mathbf{D}}^{-\frac{1}{2}}\tilde{\mathbf{A}}\tilde{\mathbf{D}}^{-\frac{1}{2}} \cdot \mathbf{C} \cdot \mathbf{W}\right). \tag{3.35}$$

The above equation describes a spectral approach of graph convolution layer which analogous to a 1-hop node information aggregation. In Equation (3.35), C' is a graph convolved signal. The filer parameters matrix $\mathbf{W}$ is learned through the back-propagation of deep models.

The model uses a four-layer Graph CNN with two convolutional layers (64-dimensional output features map in each) and two fully connected layers (producing 32- and 2-dimensional output features, respectively) to predict the fake/true class probabilities. Figure 3.10 depicts a block diagram of the model. One head of graph attention was used in every convolutional layer to implement the filters together with mean-pooling for dimensionality reduction. The Scaled Exponential Linear Unit (SELU) [70] is used as nonlinearity throughout the entire network. Hinge loss was employed to train the neural network and no regularization was used with the model.

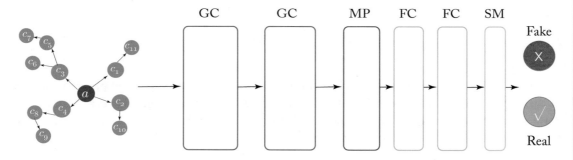

Figure 3.10: The architecture of the graph convolutional network (GCN) framework on modeling propagation network for fake news detection. GC = Graph Convolution, MP = Mean Pooling, FC = Fully Connected, SM = SoftMax layer.

3.3.6 HIERARCHICAL PROPAGATION NETWORK MODELING

In the real world, news pieces spread in networks on social media. The propagation networks have a hierarchical structure (see Figure 3.11), including macro-level and micro-level propagation networks [135]. On one hand, macro-level propagation networks demonstrate the spreading path from news to the social media posts sharing the news, and those reposts of these posts. Macro-level networks for fake news are shown to be deeper, wider, and includes more social bots than real news [127, 159], which provides clues for detecting fake news. On the other hand, micro-level propagation networks illustrate the user conversations under the posts or reposts, such as replies/comments. Micro-level networks contain user discussions toward news pieces, which brings auxiliary cues such as sentiment polarities [45] and stance signals [59] to differentiate fake news. Studying macro-level and micro-level propagation network provides fine-grained social signals to understand fake news and can possibly facilitate fake news detection.

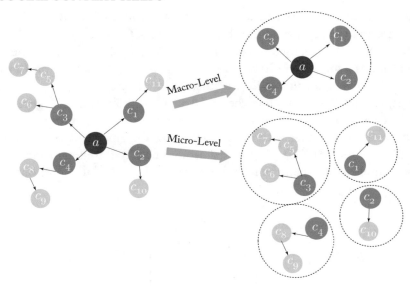

Figure 3.11: An example of the hierarchical propagation network of a fake news piece. It consists of two types: **macro-level** and **micro-level**. The macro-level propagation network includes the news nodes, tweet nodes, and retweet nodes. The micro-level propagation network indicates the conversation tree represented by reply nodes.

Macro-Level Propagation Network   Macro-level propagation network encompasses information on tweets posting pattern and information sharing pattern. We analyze the macro-level propagation network in terms of structure and temporal perspectives. Since the same textual information related to a news article is shared across the macro-level network, linguistic analysis is not applicable.

Structural analysis of macro-level networks helps to understand the global spreading pattern of the news pieces. Existing work has shown that learning latent features from the macro-level propagation paths can help to improve fake news detection, while lacking of an in-depth understanding of why and how it is helpful [80, 169]. Thus, we characterize and compare the macro-level propagation networks by looking at various network features as follows.

- (S_1) *Tree depth*: The depth of the macro propagation network, capturing how far the information is spread/retweeted by users in social media.

- (S_2) *Number of nodes*: The number of nodes in a macro network indicates the number of users who share the new article and can be a signal for understanding the spreading pattern.

- (S_3) *Maximum Outdegree*: Maximum outdegree in macro network could reveal the tweet/retweet with the most influence in the propagation process.

- **(S_4)** *Number of cascades*: The number of original tweets posting the original news article.

- **(S_5)** *Depth of node with maximum outdegree*: The depth at which node with maximum outdegree occurs. This indicates steps of propagation it takes for a news piece to be spread by an influential node whose post is retweeted by more users than any other user's repost.

- **(S_6)** *Number of cascades with retweets*: It indicate number of cascades (tweets) those were retweeted at least once.

- **(S_7)** *Fraction of cascades with retweets*: It indicates the fraction of tweets with retweets among all the cascades.

- **(S_8)** *Number of bot users retweeting*: This feature captures the number of bot users who retweet the corresponding news pieces.

- **(S_9)** *Fraction of bot users retweeting*: It is the ratio of bot users among all the users who tweeting and retweeting a news piece. This feature can show whether news pieces are more likely to be disseminated by bots or real humans.

Temporal analysis in macro-level network reveal the frequency and intensity of news dissemination process. The frequency distribution of user posting over time can be encoded in recurrent neural networks to learn the features to detection fake news [119, 133]. However, the learned features are not interpretable, and the explanation of why the learned features can help remain unclear. Here, we extract several temporal features from macro-level propagation networks explicitly for more explainable abilities and analyze whether these features are different or not. The following are the features we extracted from the macro propagation network.

- **(T_1)** *Average time difference between the adjacent retweet nodes*: It indicates how fast the tweets are retweeted in news dissemination process.

- **(T_2)** *Time difference between the first tweet and the last retweets*: It captures the life span of the news spread process.

- **(T_3)** *Time difference between the first tweet and the tweet with maximum outdegree*: Tweets with maximum outdegree in propagation network represent the most influential node. This feature demonstrates how long it took for a news article to be retweeted by most influential node.

- **(T_4)** *Time difference between the first and last tweet posting news*: This indicates how long the tweets related to a news article are posted in Twitter.

- **(T_5)** *Time difference between the tweet posting news and last retweet node in deepest cascade*: Deepest cascade represents the most propagated network in the entire propagation network. This time difference indicates the lifespan of the news in the deepest cascade and can show whether news grows in a burst or slow manner.

- **(T_6)** *Average time difference between the adjacent retweet nodes in the deepest cascade*: This feature indicates how frequent a news article is retweeted in the deepest cascade.

- **(T_7)** *Average time between tweets posting news*: This time indicates whether tweets are posted in short intervals related to a news article.

- **(T_8)** *Average time difference between the tweet post time and the first retweet time*: The average time difference between the first tweets and the first retweet node in each cascade can indicate how soon the tweets are retweeted.

Micro-Level Propagation Network   Micro-level propagation networks involve users conversations toward news pieces on social media over time. It contains rich information of user opinions toward news pieces. Next, we introduce how to extract features from micro-level propagation networks from structural, temporal, and linguistic perspectives.

Structure analysis in the micro network involves identifying structural patterns in conversation threads of users who express their viewpoints on tweets posted related to news articles.

- **(S_{10})** *Tree depth*: Depth of the micro propagation network captures how far is the conversation tree for the tweets/retweets spreading a news piece.

- **(S_{11})** *Number of nodes*: The number of nodes in the micro-level propagation network indicates the number of comments that are involved. It can measure how popular of the tweet in the root.

- **(S_{12})** *Maximum Outdegree*: In micro-network, the maximum outdegree indicates the maximum number of new comments in the chain starting from a particular reply node.

- **(S_{13})** *Number of cascade with with micro-level networks*: This feature indicates the number of cascades that have at least one reply.

- **(S_{14})** *Fraction of cascades with micro-level networks*: This feature indicates the fraction of the cascades that have at least one replies among all cascades.

Temporal analysis of micro-level propagation network depicts users' opinions and emotions through a chain of replies over time. The temporal features extracted from micro network can help understand exchange of opinions in terms of time. The following are some of the features extracted from the micro propagation network.

- **(T_9)** *Average time difference between adjacent replies in cascade*: It indicates how frequent users reply to one another.

- **(T_{10})** *Time difference between the first tweet posting news and first reply node*: It indicates how soon the first reply is posted in response to a tweet posting news.

- $(\mathbf{T_{11}})$ *Time difference between the first tweet posting news and last reply node in micro network*: It indicates how long a conversation tree lasts starting from the tweet/retweet posting a new piece.

- $(\mathbf{T_{12}})$ *Average time difference between replies in the deepest cascade*: It indicates how frequent users reply to one another in the deepest cascade.

- $(\mathbf{T_{13}})$ *Time difference between first tweet posting news and last reply node in the deepest cascade*: Indicates the life span of the conversation thread in the deepest cascade of the micro network.

Linguistic analysis people express their emotions or opinions toward fake news through social media posts, such as skeptical opinions, sensational reactions, etc. These textual information has been shown to be related to the content of original news pieces. Thus, it is necessary to extract linguistic-based features to help find potential fake news via reactions from the general public as expressed in comments from micro-level propagation network. Next, we demonstrate the sentiment features extracted from the comment posts, as the representative of linguistic features. We utilize the widely used pre-trained model VADER [45] to predict the sentiment score for each user reply, and extract a set of features related to sentiment as follows.

- $(\mathbf{L_1})$ *Sentiment ratio*: We consider a ratio of the number of replies with a positive sentiment to the number of replies with negative sentiment as a feature for each news articles because it helps to understand whether fake news gets more number of positive or negative comments.

- $(\mathbf{L_2})$ *Average sentiment*: Average sentiment scores of the nodes in the micro propagation network. Sentiment ratio does not capture the relative difference in the scores of the sentiment and hence average sentiment is used.

- $(\mathbf{L_3})$ *Average sentiment of first level replies*: This indicates whether people post positive or negative comments on the immediate tweets posts sharing fake and real news.

- $(\mathbf{L_4})$ *Average sentiment of replies in deepest cascade*: Deepest cascade generally indicate the nodes that are most propagated cascade in the entire propagation network. The average sentiment of the replies in the deepest cascade capture the emotion of user comments in most influential information cascade.

- $(\mathbf{L_5})$ *Sentiment of first level reply in the deepest cascade*: Deepest cascade generally indicate the nodes that are most propagated cascade in the entire propagation network. The sentiment of the first level reply indicates the user emotions to most influential information cascade.

We can compare all the aforementioned features for fake and real news pieces, and observe that most of the feature distributions are different. In [135], we build different learning algorithms using the extracted features to detect fake news. We evaluate the effectiveness of the extracted features by comparing with several existing baselines. The experiments show that: (1) these features can make significant contributions to help detect fake news; (2) these features are overall robust to different learning algorithms; and (3) temporal features are more discriminative than linguistic and structural features and macro- and micro-level features are complimentary.

CHAPTER 4

Challenging Problems of Fake News Detection

In previous chapters, we introduce how to extract features and build machine learning models from news content and social context to detect fake news, which generally considers the standard scenario of binary classification. Since fake news detection is a critical real-world problem, it has encountered specific challenges that need to be considered. In addition, recent advancements of machine learning methods such as deep neural networks, tensor factorization, and probabilistic models allow us to better capture effective features of news from auxiliary information, and deal with specialized settings of fake news detection.

In this chapter, we discuss several challenging problems of fake news detection. Specifically, there is a need for detecting fake news at the early stage to prevent further propagation of fake news on social media. Since obtaining the ground truth of fake news is labor intensive and time consuming, it is important to study fake news detection in a weakly supervised setting, i.e., with limited or no labels for training. It is also necessary to understand why a particular piece of news is classified as fake by machine learning models, in which the derived explanation can provide new insights and knowledge not obvious to practitioners.

4.1 FAKE NEWS EARLY DETECTION

Fake news early detection aims to give early alerts of fake news during the dissemination process so that actions can be taken to help prevent its further propagation on social media.

4.1.1 A USER-RESPONSE GENERATION APPROACH

We learn that the rich social context information provides effective auxiliary signals to advance fake news detection on social media. However, these types of social context information, such as user comments, can only be available after people have already engaged in the fake news propagation. Therefore, a more practical solution for early fake news detection is to assume the news content is the *only* available information. In addition, we can assume we have obtained historical data that contains both news contents and user response, and can leverage the historical data to help enhance early detection performance on newly emerging news articles without any user responses.

Let $\mathcal{A} = \{a_1, a_2, \cdots, a_N\}$ denote the set of news corpus, where each document a_i is a vector of term in a dictionary, Σ with size of $d = |\Sigma|$, and $\mathbf{C} = \{\mathbf{c}_1, \mathbf{c}_2, \cdots, \mathbf{c}_n\}$ represents the set of user responses. The detection task can be defined as: given a news article a, the goal is to predict whether it is fake or not without using assuming the corresponding user response exists.

The framework mainly consists of two major components (see Figure 4.1): (1) a convolution neural network component to learn news representation; and (2) a user response generator to generate auxiliary signals to help detect fake news.

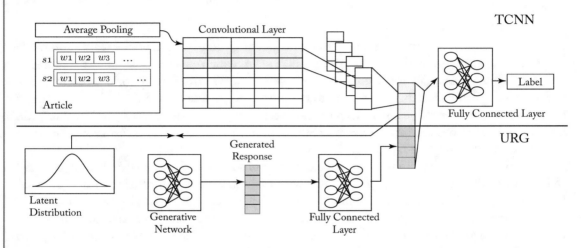

Figure 4.1: The user-response generation framework for early fake news detection. It consists primarily of two stages: neural news representation learning and deep user response generator. Based on [112].

Neural Representation Learning for News

To extract semantic information and learn the representation for news, we can use a two-level convolution neural network (TCNN) structure: sentence-level and document-level. We have introduced similar feature learning techniques in Section 2.1.3. For the sentence-level, we can first derive sentence representation as the average of the word embeddings of those words in the sentence. Each sentence in a news article can be represented as a one-hot vector $\mathbf{s} \in \{0, 1\}^{|T|}$ indicating which words from the vocabulary Σ are present in the sentence. Then the sentence representation is defined by average pooling of word embedding vectors of words in the sentence as follows:

$$\mathbf{v}(\mathbf{s}) = \frac{\mathbf{W}\mathbf{s}}{\sum_k s_k}, \tag{4.1}$$

where $\mathbf{W}$ is the embedding matrix for all words, where embedding of each word is obtained from a pre-trained skip-gram algorithm [90] on all news articles. In the document level, the news representation is derived from the sentence representations by concatenating ($\oplus$) each sentence representation. For a news piece a_i, containing L sentences $\mathcal{S} = \{s_1, \cdots, s_L\}$, the news representation $\mathbf{s}_i$ is represented as:

$$\mathbf{a}_i = \mathbf{v}(\mathbf{s}_1) \oplus \mathbf{v}(\mathbf{s}_2) \oplus \cdots \mathbf{v}(\mathbf{s}_L). \tag{4.2}$$

After the news representation is obtained, we can use the convolution neural networks to learn the representations as in Section 2.1.3.

User Response Generator

The goal of the user response generator is to generate user responses to help learn more effective representations to predict fake news. We can use a generative conditional variational auto-encoder (CVAE) [144] to learn the a distribution over user responses, conditioned on the article, and can therefore be used to generate varying responses sampled from the learned distribution. Specifically, CVAE takes the user response $\mathbf{C}^{(i)} = \{\mathbf{c}_{i1}, \cdots, \mathbf{c}_{in}\}$ and the news article $\mathbf{a}$ as the input, and aim to reconstruct the input $\mathbf{c}$ conditioned on $\mathbf{a}$ by learning the latent representation $\mathbf{z}$. The objective is shown as follows:

$$\mathbb{E}_{\mathbf{z} \sim q_\phi(\mathbf{c}_{ij}, \mathbf{a}_i)}[-\log p_\theta(\mathbf{c}_{ij}|\mathbf{z}, \mathbf{a}_i)] + D_{KL}(q_\theta(\mathbf{z}|\mathbf{c}_{ij}, \mathbf{a}_i)). \tag{4.3}$$

The first term is the reconstruction error designed as the negative log-likelihood of the data reconstructed from the latent variable $\mathbf{z}$ under the influence of article $\mathbf{a}_i$. The second term is for regularization and minimize the divergence between the encoder distribution $q_\theta(\mathbf{z}|\mathbf{c}, \mathbf{a})$ and the prior distribution $p_\theta(\mathbf{z})$.

We use the learned representation $\mathbf{a}$ from the TCNN as the condition and feed into the user response generator to generate synthetic responses. The user response generated by URG is put through a nonlinear neural network and then combined with the text features extracted by TCNN. Then, the final feature vector is fed into a feed forward softmax classifier for classification as in Figure 4.1.

4.1.2   AN EVENT-INVARIANT ADVERSARIAL APPROACH

Most of existing fake news detection methods perform supervised learning using historical data that are collected from different news events. Actually, these methods tend to capture lots of *event-specific* features which are not shared among different news events [165]. Such event-specific features, though being able to help classify the posts on verified events, would have limit help or even hurt the detection with regard to newly emerged events. Therefore, it is important to learn *event-invariant* features that are discriminative to detect fake news from unverified events. The goal is to design an effective model to remove the nontransferable event-specific features and preserve the shared event-invariant features among all the events to improve fake news detection performance.

In [165], an event adversarial neural network (EANN) model is proposed to extract nontransferable multi-modal feature representations for fake news detection (see Figure 4.2). EANN mainly consists of three components: (1) the multi-modal feature extractor; (2) the fake news detector; and (3) the event discriminator. The multi-modal feature extractor cooperates with the fake news detector to carry out the major task of identifying false news. Simultaneously, the multi-modal feature extractor tries to fool the event discriminator to learn the event invariant representations.

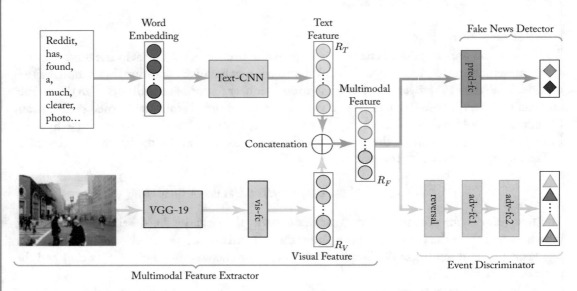

Figure 4.2: The illustration of the event adversarial neural networks (EANN). It consists of three parts: a multi-modal feature extractor, an event discriminator, and a fake news detector. Based on [165].

Multi-Modal Feature Extractor

The multi-modal feature extractor aims to extract feature representations from news text and images using neural networks. We introduced representative techniques in Sections 2.1.3 and 2.2.3 for neural textual and visual feature learning. In [165], for textual feature, the CNNs are utilized to obtain R_{cnn}; and for image feature, the VGG19 neural networks are adopted to get R_{vgg}. To enforce a standard feature representation of both text and image, we can add another dense layer ("vis-fc") to map the learned feature representation to the same dimension:

$$R_T = \sigma(\mathbf{W}_t)R_{cnn}$$
$$R_V = \sigma(\mathbf{W}_v)R_{vgg}.$$

(4.4)

The textual features R_T and the visual features R_V will be concatenated to form the multimodal feature representation denoted as $R_F = R_T \oplus R_V$, which is the output of the multimodal feature extractor.

Fake News Detector

The fake news detector deploys a fully connected layer ("pred-fc") with softmax function to predict whether a news post is fake or real. The fake news detector takes the learned multimodal feature representation R_F as the input. The objective function of fake news detector is to minimize the cross entropy loss as follows:

$$\min L_d(\theta_f, \theta_d) := \min -\mathbb{E}[y \log(P_\theta(a)) + (1 - y)(\log(1 - P_\theta(a)))], \tag{4.5}$$

where a is a news post, θ_f and θ_d are the parameters of the multi-modal feature extractor and fake news detector. However, directly optimizing the loss function in Equation (4.5) only help detect fake news on the events that are already included in the training data, so it does not generalize well to new events. Thus, we need to enable the model to learn more general feature representations that can capture the common features among all the events. Such representation should be event-invariant and does not include any event-specific features.

Event Discriminator

To learn the event-invariant feature representations, it is required to remove the uniqueness of each events in the training data and focuses on extracting features that are shared among different events. To this end, we use an event-discriminator, which is a neural network consisting of two dense layers, to correctly classify the post into one of E events $(1, \cdots, e)$ correctly. The event discriminator is a classifier and deploy the cross entropy loss function as follows:

$$\min L_e(\theta_f, \theta_e) := \min -\mathbb{E}\left[\sum_{k=1}^{E} \mathbf{1}_{[k=y_e]} \log(G_e(G_f(a; \theta_f)); \theta_e)\right], \tag{4.6}$$

where G_f denotes the multi-modal feature extractor, y_e denotes the even label predicted, and and G_e represents the event discriminator. Equation (4.6) can estimate the dissimilarities of different events' distributions. The large loss means the distributions of different events' representations are similar and the learned features are event-invariant. Thus, in order to remove the uniqueness of each event, we need to maximize the discrimination loss L_e by seeking the optimal parameter θ_f.

The above idea motivates a minimax game between the multi-modal feature extractor and the event discriminator. On one hand, the multi-modal feature extractor tries to fool the event discriminator to maximize the discrimination loss, and on the other hand, the event discriminator aims to discover the event-specific information included in the feature representations to recognize the event. The integration process of three components and the final objective function will be introduced in the next section.

4.1.3 A PROPAGATION-PATH MODELING APPROACH

The diffusion paths of fake news and real news can be very different on social media. In addition to only relying on news contents to detect fake news, the auxiliary information from spreaders at the early stage could also be important for fake news early detection. Existing detection methods mainly explore temporal-linguistic features extracted from user comments, or temporal-structural features extracted from propagation paths/trees or networks [80]. However, compared to user comments, *user characteristics* are more available, reliable, and robust in the early stage of news propagation than linguistic and structural features widely used by state-of-the-art approaches.

Given the corpus of news pieces $\mathcal{A} = \{a_1, a_2, \cdots, a_N\}$ where each document a_i is a vector of term in a dictionary, Σ with size of $d = |\Sigma|$. Let $\mathcal{U} = \{u_1, \cdots, u_n\}$ denotes the set of social media users, each user is associated with a feature vector $\mathbf{u}_i \in \mathbb{R}^k$. The propagation path is defined as a variable-length multivariate time series $\mathcal{P}(a_i) = < \cdots, (\mathbf{u}_j, t), \cdots >$, where each tuple $(\mathbf{u}_j, t)$ denotes that user u_j tweets/retweets the news story a_i. Since the goal is to perform early detection of fake news, the designed model should be able to make predictions based on only a partial propagation path, defined as $\mathcal{P}(a_i, T) = < \mathbf{x}_i, t < T >$.

This framework consists of three major components (see Figure 4.3): (1) building the propagation path; (2) learning path representations through RNN and CNN; and (3) predicting fake news base on path representations.

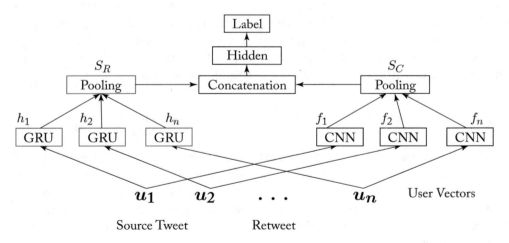

Figure 4.3: The propagation-path based framework for early fake news detection.

Building Propagation Path

The first step to build a propagation path is to identify the users who have engaged in the propagation process. The propagation path $\mathcal{P}(a_i)$ for news piece a_i is constructed by extracting the user characteristics from those profiles of the users who posts/reposts the news piece. After $\mathcal{P}(a_i)$ is obtained, the length of the propagation path would be different for different news pieces. Therefore, we can transform all the propagation paths with the fixed lengths n, denoted as $\mathcal{S}(a_i) =< \mathbf{u}_1, \cdots, \mathbf{u}_n >$. If there are more than n tuples in $\mathcal{P}(a_i)$, then $\mathbf{P}(a_i)$ will be truncated and only the first n tuples appear in $\mathcal{S}(a_i)$; if $\mathcal{P}(a_i)$ contains less than n tuples, then tuples are over-sampled from $\mathcal{P}(a_i)$ to ensure the final length of $\mathcal{S}(a_i)$ is n.

Learning Path Representations

To learn the representation, both RNNs and CNNs are utilized to preserve the global and local temporal information [80]. We have introduced how to use RNN to learn temporal representation in Section 3.3.3; a similar technique can be used here. As shown in Figure 4.3, we can obtain the representation $\mathbf{h}_t$ at each timestamp t, and the overall representation using RNN can be computed as the mean pooling of all output vectors $< \mathbf{h}_1, \cdots, \mathbf{h}_n >$ for all timestamps, i.e., $s_R = \frac{1}{n} \sum_{t=1}^{n} \mathbf{h}_t$, which encodes the global variation of user characteristics.

To encode the local temporal feature of user characteristics, we first construct the local propagation path $\mathbf{U}_{t:t+h-1} =< \mathbf{u}_t, \cdots, \mathbf{u}_{t+h-1} >$ for each timestamp t, with a moving window h out of $\mathcal{S}(a_i)$. Then we apply CNN with a convolution filter on $\mathbf{U}_{t:t+h-1}$ to get a scalar feature c_t:

$$c_t = ReLU(\mathbf{W}_f \mathbf{U}_{t:t+h-1} + \mathbf{b}_f), \tag{4.7}$$

where $ReLU$ is the dense layer with rectified linear unit activation function and $\mathbf{b}_f$ is a bias term. To map the c_t to a latent vector, we can utilize a convolution filter to transform c_t to a k dimension vector $\mathbf{f}_t$. So, we can obtain a feature vectors for all the timestamps $< \mathbf{f}_1, \cdots, \mathbf{f}_{n-h+1} >$, from which we can apply mean pooling operation and get the overall representations $s_C = \frac{1}{n} \sum_{t=1}^{n-h+1} \mathbf{f}_t$ that encodes the local representations of user characteristics.

Predicting Fake News

After we learn the representations from propagation paths through both the RNN-based and CNN-based techniques, we can concatenate them into a single vector as follows:

$$\mathbf{a} = [s_R, s_C] \tag{4.8}$$

and then $\mathbf{a}$ is fed into a multi-layer (q layer) feed-forward neural network that finally to predict the class label for the corresponding propagation path as follows:

$$\begin{aligned} \mathbf{I}_j &= ReLU(\mathbf{W}_j \mathbf{I}_{j-1} + \mathbf{b}_j), \forall j \in [1, \cdots, q] \\ \mathbf{y} &= softmax(\mathbf{I}_q), \end{aligned} \tag{4.9}$$

where q is the number of hidden layers, $\mathbf{I}_j$ is the hidden states of the j_{th} layer, and $\mathbf{y}$ is the output over the set of all possible labels of news pieces.

4.2   WEAKLY SUPERVISED FAKE NEWS DETECTION

Weakly supervised fake news detection aims to predict fake news with limited or no supervision labels. We introduce some representative methods for semi-supervised and unsupervised fake news detection.

4.2.1   A TENSOR DECOMPOSITION SEMI-SUPERVISED APPROACH

In the scenario of semi-supervised fake news detection, assuming the labels of a limited number of news pieces are given, we aim to predict the labels of unknown news articles. Formally, we denote the corpus of fake news $\mathcal{A} = \{a_1, a_2, \cdots, a_N\}$ where each document a_i can be represented as a vector of term in a dictionary, Σ with size of $d = |\Sigma|$. Assuming that the labels of some news articles are available, and let $y \in \{-1, 0, 1\}$ denote the vector containing the partially known labels, such that entries of 1 represent real news, -1 represents fake news, and 0 denotes an unknown label. The problem is that, given a collection of news articles a and a label y with entries for labeled real/fake news and unknown articles, the goal is to predict the class labels of the unknown articles.

The framework for semi-supervised fake news detection mainly consists of two stages (see Figure 4.4): (1) tensor decomposition; (2) building k-Nearest-Neighbor (KNN) news graph; and (3) belief propagation. The tensor decomposition is to learn news representation from news contents; the KNN graph is built to link labeled and unlabeled news pieces; and belief propagation can utilize graph structure modeling to predict the labels of unknown news pieces.

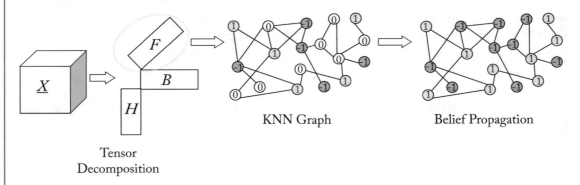

Figure 4.4: The tensor decomposition framework for semi-supervised fake news detection.

Building KNN Graph of News Articles

Spatial relations among words represent the affinities of them in a news document. To incorporate spatial relations, a news document a_i can be represented as a tensor $\underline{\mathbf{X}} \in \mathbb{R}^{N \times d \times d}$. We then decompose the three mode tensor $\underline{\mathbf{X}} \in \mathbb{R}^{N \times d \times d}$ into three matrix using the CP/PARAFAC decomposition as in Equation (2.2) in Section 2.1.2. Then we can obtain the news representation matrix $\mathbf{F}$.

The tensor embedding we computed in last step provides a compact and discriminative representation of news articles into a concise set of latent topics. Using this embedding, we construct a graphical representation of news articles. Specifically, we can used the resultant news representation matrix $\mathbf{F} \in \mathbb{R}^{N \times d}$ to build a KNN graph G among labeled and unlabeled news pieces. Each node in G represents a news article and each edge encodes that two articles are similar in the embedding space. A KNN graph is a graph where a node $\mathbf{a}_i$ and $\mathbf{a}_j$ are connected by an edge is one of them belongs to the KNN list of the other. The KNN of a data point is defined using the closeness relation in the feature space with a distance metric such as Euclidean l_2 distance as follows. In practice, the parameter K is empirically decided:

$$d(\mathbf{F}_i, \mathbf{F}_j) = \|\mathbf{F}_i - \mathbf{F}_j\|_2. \tag{4.10}$$

The resultant graph G is an undirected, symmetric graph where each node is connected to at least k nodes. The graph can be compactly represented as an $N \times N$ adjacency matrix $\mathbf{O}$.

Belief Propagation

Belief propagation is to propagate the label information of labeled news pieces to unlabeled ones in the KNN graph G. The basic assumption is that news articles that are connected in G are more likely to be of the same labels due to the construction method of the tensor embeddings. To this end, a fast and linearized fast belief propagation (FaBP) [73] can be utilized due to the efficiency and effectiveness in large graphs. The operative intuition behind FaBP and other such guilt-by-association methods is that nodes which are "close" are likely to have similar labels or belief values. The FaBP solves the following linear problem:

$$[\mathbf{I} + \alpha \mathbf{D} - c'\mathbf{O}]b_h = \phi_h, \tag{4.11}$$

where ϕ_h and b_h denote the prior and final beliefs (labels). $\mathbf{O}$ is the adjacency matrix of the KNN graph, $\mathbf{I}$ is an identity matrix, and $\mathbf{D}$ is a diagonal matrix where $\mathbf{D}_{ii} = \sum_j \mathbf{O}_{ij}$ and $\mathbf{D}_{ij} = 0$ for $i \neq j$.

4.2.2   A TENSOR DECOMPOSITION UNSUPERVISED APPROACH

The news content is suggested to contain crucial information to differentiate fake from real news [53]. Section 2.1.1 introduced linguistic textual feature extraction for fake news detection. These features such as n-gram, and tf-idf (term frequency-inverse document frequency) can capture the correlations and similarities between different news contents. However, they ignore the context of a news document such as spatial vicinity of each word. To this end, we model the corpus as a third-order tensor, which can simultaneously leverage the article and term relations, as well as the spatial/contextual relations between the terms. The advantage is that by exploiting both aspects of the corpus, in particular the spatial relations between the words, the learned news representation can be a determining factor for identifying groups of articles that fall under different types of fake news.

Given the corpus of news pieces $\mathcal{A} = \{a_1, a_2, \cdots, a_N\}$ where each document a_i is a vector of term in a dictionary, Σ with size of $d = |\Sigma|$. The problem is clustering the news documents based on their terms into different categories such as fake and real.

The framework for unsupervised fake news detection consists of two stages (see Figure 4.5). First, we explain how to extract spatial relations between terms through tensor decomposition; then, we discus the co-clustering to decompose relevant documents.

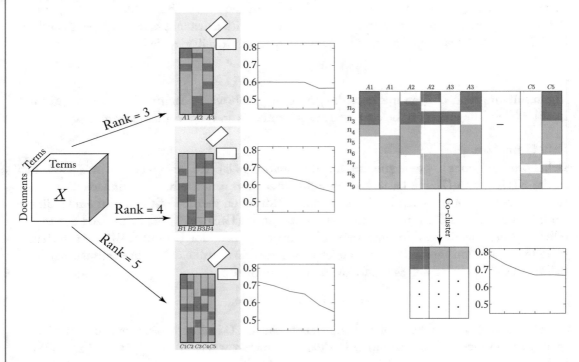

Figure 4.5: A tensor decomposition framework for unsupervised fake news detection. It consists of two stages: tensor decomposition for spatial relation extraction and co-clustering decomposition ensembles. Based on [53].

Tensor Ensemble Co-Clustering

We first build a tensor $\underline{\mathbf{X}} \in \mathbb{R}^{N \times d \times d}$ for each news a. We then decompose the tensor $\underline{X}$ into three matrix using the CP/PARAFAC decomposition as in Equation (2.2) in Section 2.1.2, and we obtain the news representation matrix $\mathbf{F}$. After learning the news representation through spatial relation extraction process, the next step is to cluster the news pieces based on their cluster membership among a set of different tensor decompositions. Intuitively, the clustering seeks to find a subset of news articles that frequently cluster together in different configurations of the tensor decomposition of the previous step. The intuition is that news articles that tend to frequently appear surrounded each other among different rank configurations, while having the same ranking within their latent factors are more likely to ultimately belong to the same category. The ranking of a news article with respect to a latent factor is derived by simply sorting the coefficients of each latent factor, corresponding to the clustering membership of a news article to a latent factor.

To this end, we can combine the clustering results of each individual tensor decomposition into a collective (news-article by latent-factor) matrix, from which we are going to extract co-clusters of news articles and the corresponding latent factors (coming from the ensemble of decompositions). For example, as shown in Figure 4.5, we can perform the tensor decomposition three times with different rank 3, 4, and 5, and then construct a collect feature matrix $\mathbf{F}'$. The co-clustering objective with l_1 norm regularization for a combine matrix [103] $\mathbf{F}'$ is shown as follows:

$$\min \left\| \mathbf{F}' - \mathbf{R}\mathbf{Q}^T \right\|_F^2 + \lambda(\|\mathbf{R}\|_1 + \|\mathbf{Q}\|_1), \qquad (4.12)$$

where $\mathbf{R} \in \mathbb{R}^{N \times k}$ is the representation matrix of news articles, and $\mathbf{Q} \in \mathbb{R}^{M \times k}$ is the coding matrix, and the term $\lambda(\|\mathbf{R}\|_1 + \|\mathbf{Q}\|_1)$ is to enforce the sparse constraints.

4.2.3 A PROBABILISTIC GENERATIVE UNSUPERVISED APPROACH

Existing work on fake news detection is mostly based on supervised methods. Although they have shown some promising results, these supervised methods suffer from a critical limitation, i.e., they require a reliably pre-annotated dataset to train a classification model. However, obtaining a large number of annotations is time consuming and labor intensive, as the process needs careful checking of news contents as well as other additional evidence such as authoritative reports.

The key idea is to extract users' opinions on the news by exploiting the auxiliary information of the users' engagements with the news tweets on social media, and aggregate their opinions in a well-designed unsupervised way to generate our estimation results [174]. As news propagates, users engage with different types of behaviors on social media, such as publishing a news tweet, liking, forwarding, or replying to a news tweet. This information can, on a certain level, reflect the users opinions on the news. For example, Figure 4.6 shows two news tweet examples regarding the aforementioned news. According to the users' tweet contexts, we see that the user in Figure 4.6a disagreed with the authenticity of the news, which may indicate the user's high credibility in identifying fake news. On the other hand, it appears that the user in Figure4.6b falsely believed the news or intentionally spread the fake news, implying the user's deficiency in the ability to identify fake news. Besides, as for other users who engaged in the tweets, it is likely that the users who liked/retweeted the first tweet also doubted the news, while those who liked/retweeted the second tweet may also be deceived by the news. The users'

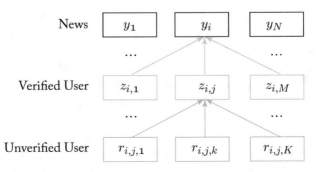

> iYamWhatIYam @MRIrene · 21 Oct 2016
> **FALSE: Pope Francis Shocks World, Endorses Donald Trump for President**
> Trumpbots getting desperate and creative. go.shr.lc/2cNK449
>
> ○   ⟲ 4   ♡ 3   ✉

(a) Doubting the authenticity of the news

> Janie Johnson ● @jjauthor · 4 Nov 2016
> Not shocking! Vote Babies!
>
> **Pope Francis Shocks World, Endorses Donald Trump for President, Releases Statement** endingthefed.com/pope-francis-s...
>
> ○ 12   ⟲ 58   ♡ 46   ✉

(b) Agreeing to the authenticity of the news

Figure 4.6: The illustration of user opinions to news pieces. Based on [174].

opinions toward the news can also be revealed from their replies to the news tweets. Next, we introduce the scenario of the hierarchical structure of user engagements, and then describe how to use a probabilistic graph model to infer fake news.

Hierarchical User Engagements   Figure 4.7 presents an overview of the hierarchical structure of user engagements on social media. Let $\mathcal{A} = \{a_1, \cdots, a_N\}$ denote a set of news corpus, and $\mathcal{U}^M$ and $\mathcal{U}^K$ represent the sets of verified and unverified users , respectively. For each given news $a_i \in \mathcal{A}$, we collect all the verified users' tweets on this news. Let $\mathcal{U}_i^M \in \mathcal{U}^M$ denote the set of verified users who published tweets for the news a_i. Then, for the tweet of each verified user $u_j \in \mathcal{U}_{Mi}$, we collect the unverified users' social engagements. Let $\mathcal{U}_{ij}^K \in \mathcal{U}^K$ denote the set of unverified users who engaged in the tweet.

News	y_1	y_i	y_N
	...		...
Verified User	$z_{i,1}$	$z_{i,j}$	$z_{i,M}$
	...		...
Unverified User	$r_{i,j,1}$	$r_{i,j,k}$	$r_{i,j,K}$

Figure 4.7: An illustration of hierarchical user engagements.

For each verified user $u_j \in \mathcal{U}_i^M$, we let $z_{i,j} \in \{0, 1\}$ denote the user's opinion on the news, i.e., $z_{i,j}$ is 1 if the user thinks the news is real, and 0 otherwise. Several heuristics can be applied to extract $z_{i,j}$. Let a_i and $c_{i,j}$ denote the news content and the user u_j's own text content of the tweet, respectively. Then, $z_{i,j}$ can be defined as the sentiment of $c_{i,j}$.

For verified user u_j's tweet on news a_i, many unverified users may like, retweet, or reply to the tweet. Let $r_{i,j,k} \in \{0, 1\}$ denote the opinion of the unverified user $u_k \in \mathcal{U}_{ij}^K$. We assume that if the user u_k liked or retweeted the tweet, then it implies that u_k agreed to the opinion of the tweet. If the user u_k replied to the tweet, then its opinion can be extracted by employing off-the-shelf sentiment analysis [55]. When an unverified user may conduct multiple engagements in a tweet, the user's opinion $r_{i,j,k}$ is obtained using majority voting.

Probabilistic Graphic Model for Inferring Fake News   Given the definitions of a_i, $z_{i,j}$, and $r_{i,j,k}$, we now present the unsupervised fake news detection framework (UFD). Figure 4.8 shows the probabilistic graphical structure of the model. Each node in the graph represents a random variable or a prior parameter, where darker nodes and white nodes indicate observed or latent variables, respectively.

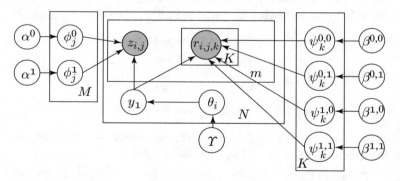

Figure 4.8: A probabilistic graphic model for unsupervised fake news detection. The circles with gray colors denote observed random variables, the circles without white color are non-observed random variables, arrows mean the conditional dependencies, and rectangles illustrate the duplication frequencies.

Each **news** a_i, y_i is generated from a Bernoulli distribution with parameter θ_i:

$$y_i \sim Bernoulli(\theta_i). \tag{4.13}$$

The prior probability of θ_i is generated from a Beta distribution with hyper parameter $\gamma = (\gamma_1, \gamma_0)$ as $\theta_i \sim Beta(\gamma_1, \gamma_0)$, where γ_1 is the prior number of true news pieces and γ_0 is the prior number of fake news pieces. If we do not have a strong belief in practice, we can initially assign a uniform prior indicating that each news piece has an equal probability of being true or fake.

For **verified user** u_j, its credibility toward news is modeled with two variables ϕ_j^1 and ϕ_j^0, denoting the probability that user u_j thinks a news piece is real giving the true estimation of the news is true or fake, defined as follows:

$$\phi_j^1 := p(z_{i,j} = 1 | y_i = 1)$$
$$\phi_j^0 := p(z_{i,j} = 1 | y_i = 0). \tag{4.14}$$

We generate the parameters ϕ_j^1 from Beta distributions with hyper parameters $\boldsymbol{\alpha}^1 = (\alpha_1^1, \alpha_0^1)$, where α_1^1 is the prior true positive count, and α_0^1 denotes the false negative count. Similarly, we can generate ϕ_j^0 from another Beta distribution with hyper parameters $\boldsymbol{\alpha}^0 = (\alpha_1^0, \alpha_0^0)$:

$$\phi_j^1 \sim \text{Beta}\left(\alpha_1^1, \alpha_0^1\right)$$
$$\phi_j^0 \sim \text{Beta}\left(\alpha_1^0, \alpha_0^0\right). \tag{4.15}$$

Given ϕ_j^1 and ϕ_j^0, the opinion of each verified user u_j for news i is generated from a Bernoulli distribution with parameter $\phi_j^{x_i}$, $y_{i,j} \sim \textit{Bernoulli}(\phi_j^{x_i})$.

For **unverified user**, he/she engages in the verified users' tweets, and thus the opinion is likely to be influenced by the news itself and the verified users' opinions. Therefore, for each unverified user $u_k \in \mathcal{U}^K$, the following variables are adopted to model the credibility:

$$\psi_k^{0,0} := p(r_{i,j,k} = 1 | x_i = 0, z_{i,j} = 0)$$
$$\psi_k^{0,1} := p(r_{i,j,k} = 1 | x_i = 0, z_{i,j} = 1)$$
$$\psi_k^{1,0} := p(r_{i,j,k} = 1 | x_i = 1, z_{i,j} = 0)$$
$$\psi_k^{1,1} := p(r_{i,j,k} = 1 | x_i = 1, z_{i,j} = 1) \tag{4.16}$$

and for each pair of $(u, v) \in \{0, 1\}^2$, $\psi_k^{u,v}$ represents the probability that the unverified user u_k thinks the news is true under the condition that the true labels of the news is u and the verified user' opinion is v. For each $\psi_k^{u,v}$, it is generated from the Beta distribution as follows:

$$\phi_k^{u,v} \sim \text{Beta}\left(\beta_1^{u,v}, \beta_0^{u,v}\right). \tag{4.17}$$

Given the truth estimation of news y_i, and the verified user' opinions $z_{i,j}$, we generate the unverified user's opinion from a Bernoulli distribution with parameter $\phi_k^{y_i, z_{i,j}}$ as follows:

$$r_{i,j,k} = \text{Bernoulli}\left(\phi_k^{y_i, z_{i,j}}\right). \tag{4.18}$$

The overall objective is to find instances of the latent variables that maximize the joint probability estimation of $\mathbf{y}$ as follows:

$$\hat{\mathbf{y}} = \text{argmax}_{\mathbf{y}} \iiint p(\mathbf{y}, \mathbf{z}, \mathbf{r}, \boldsymbol{\theta}, \boldsymbol{\phi}, \boldsymbol{\psi}) d\boldsymbol{\theta} \, d\boldsymbol{\phi} \, d\boldsymbol{\psi}, \tag{4.19}$$

where for simplicity of notations, we use $\boldsymbol{\phi}$ and $\boldsymbol{\psi}$ to denote $\{\phi^0, \phi^1\}$ and $\{\psi^{0,0}, \psi^{0,1}, \psi^{1,0}, \psi^{1,1}\}$, respectively. However, the exact inference on the posterior distribution may result in exponential complexity; we propose using Gibbs sampling to effectively inference the variable estimations.

4.3 EXPLAINABLE FAKE NEWS DETECTION

In recent years, computational detection of fake news has been producing some promising early results. However, there is a critical missing piece of the study, the explainability of such detection, i.e., *why* a particular piece of news is *detected* as fake. Here, we introduce two representative approaches based on web articles and user comments.

4.3.1 A WEB EVIDENCE-AWARE APPROACH

Web evidence is important to provide additional auxiliary information to predict the credibility of online misinformation and fake news. Existing methods for fake news detection focus on exploring effective features from different sources such as the news content or social media signals to improve fake news detection performance. However, these approaches also do not offer any explanation of their verdicts. In the real world, external evidence or counter-evidence from the Web can serve as a base to mine user-unrepeatable explanations.

Given a set of n news claims $\mathcal{A} = \{a_1, \cdots, a_N\}$ with their corresponding sources $\mathcal{P} = \{p_1, \cdots, p_N\}$, and each news claim a_i is reported by a set of L articles $\mathcal{W}_i = \{e_{i,1}, \cdots, e_{i,L}\}$, where $i \in [1, N]$, from sources $\mathcal{WP}_i = \{ep_{i,1}, \cdots, ep_{i,L}\}$. Each tuple $(a_i, p_i, e_{ij}, ep_{ij})$ forms a training instance. The goal is to predict the label for each news claim as fake or real, with user-comprehensible explanations for the prediction results (see Figure 4.9).

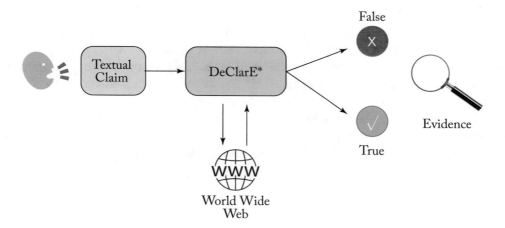

Figure 4.9: The illustration of fake news detection with evidence-aware explanations.

The framework *DeClarE*[110] (see Figure 4.10), debunking claims with interpretable evidence, mainly consists of the following components: (1) learning Claims and Article Representations; (2) claim Specific Attention; and (3) claim Credibility Prediction.

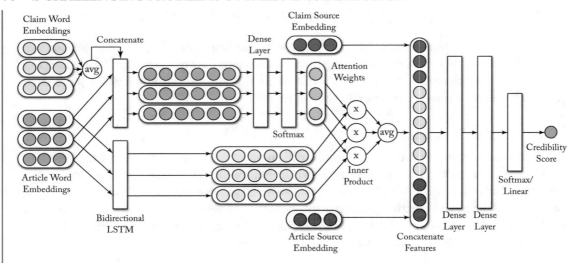

Figure 4.10: The framework of fake news detection with evidence-aware explanations. Based on [110].

Learning Claims and Article Representations

The claim $\mathbf{a}_i$ of length l is represented as $[\mathbf{w}_1, \mathbf{w}_2, \cdots, \mathbf{w}_l]$ where $\mathbf{w}_l$ is the word embedding vectors for the l-th word. The source of the claim and articles are represented by embedding vectors with the same dimensions. A report article from the Web $\mathbf{e}_{i,j}$ is represented by $[\mathbf{w}_{i,j,1}, \mathbf{w}_{i,j,2}, \cdots, \mathbf{w}_{i,j,k}]$, where $\mathbf{w}_{i,j,k}$ is the word embedding vector of the k-th word in the article. To obtain the representation of an article e_{ij}, we can use the bidirectional LSTM networks, as we introduced in Section 2.1.3, to learn the neural textual representations. Specifically, given an input word embedding of token w_k, an LSTM cell performs nonlinear transformations to generate a hidden state $\mathbf{h}_k$ for timestamp k, and the last hidden state can be regard as the representation. Since bidirectional LSTM is adopted, so the final out representation is the concatenation of the output of forward LSTM and backward LSTM, i.e., $\mathbf{h} = [\overrightarrow{\mathbf{h}}, \overleftarrow{\mathbf{h}}]$.

Claim-Specific Attention

To consider the relevance of an article with respect to the claim, we can use attention mechanism to help the model focus on salient words in the article. By adding attention mechanism, it also helps make the model transparent and interpretable. First, the overall representation of the input claim is generated by taking the average of the word embeddings of all words:

$$\bar{\mathbf{a}} = \frac{1}{l} \sum_l \mathbf{w}_l \tag{4.20}$$

and then the overall representation of this claim $\mathbf{a}$ is concatenated with each article as follows:

$$\hat{\mathbf{a}}_k = \mathbf{e}_k \oplus \bar{\mathbf{a}} \tag{4.21}$$

the claim-specific representation of each article is transformed through a fully connected layer:

$$\mathbf{a}'_k = f(\mathbf{W}_a \hat{\mathbf{a}}_k + \mathbf{b}_a), \tag{4.22}$$

where $\mathbf{W}_a$ and $\mathbf{b}_a$ are the corresponding weight matrix and bias terms, and f is the activation function. Following this step, a softmax activation can be used to calculate an attention score α_k for each word in the article capturing the relevance to the claim context:

$$\alpha_k = \frac{\exp(\mathbf{a}'_k)}{\sum_k \exp(\mathbf{a}'_k)}. \tag{4.23}$$

Now that we have the article representation $< \mathbf{h}_k >$, and their relevance to the claim given by $< \alpha_k >$, we can combine then to further predict the news claims' credibility. The weighted average of the hidden state representations for all articles can be calculated as follows:

$$\mathbf{r} = \frac{1}{k} \sum_k \alpha_k \cdot \mathbf{h}_k. \tag{4.24}$$

At last, the article representation $\mathbf{r}$ is combined with the claim source embedding (ep) and article source embedding (p) simultaneously through a fully connected layer,

$$\mathbf{l} = relu(\mathbf{W}_c(\mathbf{r} \oplus \mathbf{ep} \oplus \mathbf{p}) + \mathbf{b}_c). \tag{4.25}$$

Claim Credibility Prediction

The credibility score for each article y is predicted by taking the aforementioned representation into a softmax layer:

$$e = softmax(\mathbf{l}). \tag{4.26}$$

Therefore, once we have the per-article credibility scores, we can take the average of these scores to generate the overall credibility score for the news claim:

$$y = \frac{1}{L} \sum_j e_j. \tag{4.27}$$

4.3.2   A SOCIAL CONTEXT-AWARE APPROACH

One way is to derive explanation from the perspectives of news contents and user comments (see Figure 4.11) [132]. First, news contents may contain information that is verifiably false. For example, journalists manually check the claims in news articles on fact-checking websites such as PolitiFact,[1] which is usually labor intensive and time consuming. Researchers also attempt to use external sources to fact-check the claims in news articles to decide and explain whether a news piece is fake or not [29], which may not be able to check newly emerging events (that has not been fact-checked). Second, user comments have rich information from the crowd on social media, including opinions, stances, and sentiment, that are useful to detect fake news. For example, researchers propose to use social features to select important comments to predict fake news pieces [48]. Moreover, news contents and user comments inherently are *related* each other and can provide important cues to explain why a given news article is fake or not. For example, in Figure 4.11, we can see users discuss different aspects of the news in comments such as "St. Nicholas was white? Really??Lol," which directly responds to the claims in the news content "The Holy Book always said Santa Claus was white."

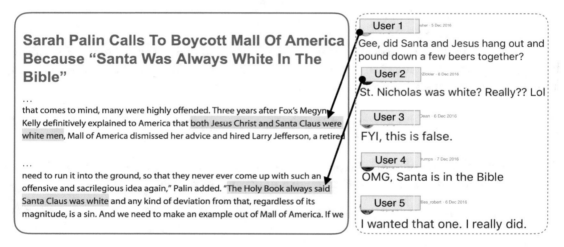

Figure 4.11: A piece of fake news on PolitiFact, and the user comments on social media. Some explainable comments are directly related to the sentences in news contents.

Let a be a news article, consisting of L sentences $\{s_i\}_{i=1}^{L}$. Each sentence $s_i = \{w_1^i, \cdots, w_{M_i}^i\}$ contains M_i words. Let $C = \{c_1, c_2, ..., c_T\}$ be a set of T comments related to the news a, where each comment $c_j = \{w_1^j, \cdots, w_{Q_j}^j\}$ contains Q_j words. Similar to previous research [59, 136], we treat fake news detection problem as the binary classification problem, i.e., each news article can be true ($y = 1$) or fake ($y = 0$). At the same time, we aim to learn a rank list RS from all sentences in $\{s_i\}_{i=1}^{L}$, and a rank list RC from all comments in $\{c_j\}_{j=1}^{T}$, according to the degree of explainability, where RS_k (RC_k) denotes the k_{th} most explainable sentence (comment). The explainability of sentences in news contents represent the degree of how check-worthy they are, while the explainability of comments denote the degree of how much users believe if news is fake or real, closely related to the major claims in news.

[1]https://www.politifact.com/

We present the details of the framework for explainability fake news detection, named as dEFEND (neural Explainable FakE News Detection). It consists of four major components (see Figure 4.12): (1) a news content encoder (including word encoder and sentence encoder) component; (2) a user comment encoder component; (3) a sentence-comment co-attention component; and (4) a fake news prediction component.

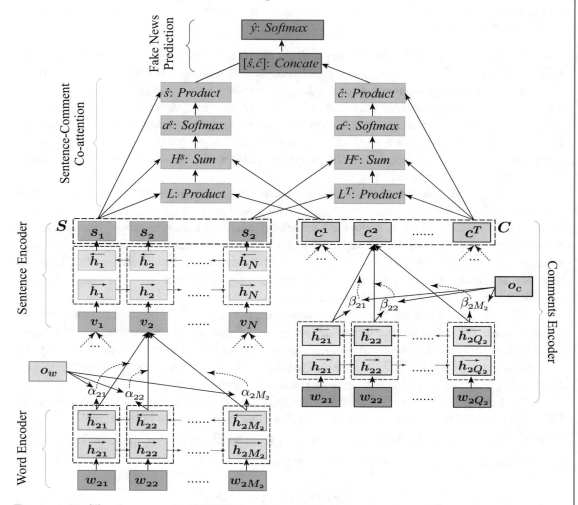

Figure 4.12: The framework dEFEND consists of four components: (1) a news content (including word-level and sentence-level) encoder; (2) a user comment encoder; (3) a sentence-comment co-attention component; and (4) a fake news prediction component. Based on [132].

First, the news content encoder component describes the modeling from the news linguistic features to latent feature space through a hierarchical word- and sentence-level encoding; next, the user comment encoder component illustrates the comment latent feature extraction through word-level attention networks; then, the sentence-comment co-attention component models the mutual influences between the news sentences and user comments for learning feature representations, and the explainability degree of sentences and comments are learned through the attention weights within co-attention learning; finally, the fake news prediction component shows the process of concatenating news content and user comment features for fake news classification.

News Contents Encoding   As fake news pieces are intentionally created to spread inaccurate information rather than to report objective claims, they often have opinionated and sensational language styles, which have the potential to help detect fake news. In addition, a news document contains linguistic cues with different levels such as word-level and sentence-level, which provide different degrees of importance for the explainability of why the news is fake. For example, in a fake news claim "Pence: Michelle Obama is the most vulgar first lady we've ever had," the word "vulgar" contributes more signals to decide whether the news claim is fake rather than other words in the sentence.

Recently, researchers find that hierarchical attention neural networks [177] are very practical and useful to learn document representations [24] with highlighting important words or sentences for classification. It adopts a hierarchical neural network to model word-level and sentence-level representations through self-attention mechanisms. Inspired by [24], we learn the news content representations through a hierarchical structure. Specifically, we first learn the sentence vectors by using the word encoder with attention and then learn the sentence representations through sentence encoder component.

Word Encoder   We learn the sentence representation via a RNN based word encoder. Although in theory, RNN is able to capture long-term dependency, in practice, the old memory will fade away as the sequence becomes longer. To making it easier for RNNs to capture long-term dependencies, GRU [27] are designed in a manner to have more persistent memory. Similar to [177], we adopt GRU to encode the word sequence.

To further capture the contextual information of annotations, we use bidirectional GRU [8] to model word sequences from both directions of words. The bidirectional GRU contains the forward GRU $\overrightarrow{f}$ which reads sentence s_i from word w_{i1} to w_{iM_i} and a backward GRU $\overleftarrow{f}$ which reads sentence s_i from word w_{iM_i} to w_{i1}:

$$
\begin{aligned}
\overrightarrow{\mathbf{h}_{it}} &= \overrightarrow{GRU}(\mathbf{w}_{it}), t \in \{1, \dots, M_i\} \\
\overleftarrow{\mathbf{h}_{it}} &= \overleftarrow{GRU}(\mathbf{w}_{it}), t \in \{M_i, \dots, 1\}.
\end{aligned}
\tag{4.28}
$$

We then obtain an annotation of word w_{it} by concatenating the forward hidden state $\overrightarrow{\mathbf{h}_{it}}$ and backward hidden state $\overleftarrow{\mathbf{h}_{it}}$, i.e., $\mathbf{h}_{it} = [\overrightarrow{\mathbf{h}_{it}}, \overleftarrow{\mathbf{h}_{it}}]$, which contains the information of the whole sentence centered around w_{it}.

Note that not all words contribute equally to the representation of the sentence meaning. Therefore, we introduce an attention mechanism to learn the weights to measure the importance of each word, and the sentence vector $\mathbf{v}_i \in \mathbb{R}^{2d \times 1}$ is computed as follows:

$$\mathbf{v}_i = \sum_{t=1}^{M_i} \alpha_{it} \mathbf{h}_{it}, \qquad (4.29)$$

where α_{it} measures the importance of tth word for the sentence s_i, and α_{it} is calculated as follows:

$$\mathbf{o}_{it} = \tanh(\mathbf{W}_w \mathbf{h}_{it} + \mathbf{b}_w)$$
$$\alpha_{it} = \frac{\exp(\mathbf{o}_{it}\mathbf{o}_w^\mathsf{T})}{\sum_{k=1}^{M_i} \exp(\mathbf{o}_{ik}\mathbf{o}_w^\mathsf{T})}, \qquad (4.30)$$

where $\mathbf{o}_{it}$ is a hidden representation of $\mathbf{h}_{it}$ obtained by feeding the hidden state $\mathbf{h}_{it}$ to a fully embedding layer, and $\mathbf{o}_w$ is the weight parameter that represents the world-level context vector.

Sentence Encoder Similar to word encoder, we utilize RNNs with GRU units to encode each sentence in news articles. Through the sentence encoder, we can capture the context information in the sentence-level to learn the sentence representations $\mathbf{h}_i$ from the learned sentence vector $\mathbf{v}_i$. Specifically, we can use the bidirectional GRU to encode the sentences as follows:

$$\overrightarrow{\mathbf{h}_i} = \overrightarrow{GRU}(\mathbf{v}_i), i \in \{1, \ldots, L\}$$
$$\overleftarrow{\mathbf{h}_i} = \overleftarrow{GRU}(\mathbf{v}_i), i \in \{L, \ldots, 1\}. \qquad (4.31)$$

We then obtain an annotation of sentence $s_i \in \mathbb{R}^{2d \times 1}$ by concatenating the forward hidden state $\overrightarrow{\mathbf{h}_i}$ and backward hidden state $\overleftarrow{\mathbf{h}_i}$, i.e., $s_i = [\overrightarrow{\mathbf{h}_i}, \overleftarrow{\mathbf{h}_i}]$, which captures the context from neighbor sentences around sentence s_i.

User Comments Encoding   People express their emotions or opinions toward fake news through social media posts such as comments, such as skeptical opinions, sensational reactions, etc. These textual information has been shown to be related to the content of original news pieces. Thus, comments may contain useful semantic information that has the potential to help fake news detection. Next, we demonstrate how to encode the comments to learn the latent representations. The comments extracted from social media are usually short text, so we use RNNs to encode the word sequence in comments directly to learn the latent representations of comments. Similar to the word encoder, we adopt bidirectional GRU to model the word sequences in comments. Specifically, given a comment c_j with words $w_{jt}, t \in \{1, \cdots, Q_j\}$, we first map each word w_{jt} into the word vector $\mathbf{w}_{jt} \in \mathbb{R}^d$ with an embedding matrix. Then, we can obtain the feed forward hidden states $\overrightarrow{\mathbf{h}_{jt}}$ and backward hidden states $\overleftarrow{\mathbf{h}_{jt}}$ as follows:

$$\overrightarrow{\mathbf{h}_{jt}} = \overrightarrow{GRU}(\mathbf{w}_{jt}), t \in \{1, \ldots, Q_j\}$$
$$\overleftarrow{\mathbf{h}_{jt}} = \overleftarrow{GRU}(\mathbf{w}_{jt}), t \in \{Q_j, \ldots, 1\}. \qquad (4.32)$$

We further obtain the annotation of word w_{jt} by concatenating $\overrightarrow{\mathbf{h}_{jt}}$ and $\overleftarrow{\mathbf{h}_{jt}}$, i.e., $\mathbf{h}_{jt} = [\overrightarrow{\mathbf{h}_{jt}}, \overleftarrow{\mathbf{h}_{jt}}]$. We also introduce the attention mechanism to learn the weights to measure the importance of each word, and the comment vector $\mathbf{c}_j \in \mathbb{R}^{2d}$ is computed as follows:

$$\mathbf{c}_j = \sum_{t=1}^{Q_j} \beta_{jt} \mathbf{h}_{jt}, \tag{4.33}$$

where β_{jt} measures the importance of tth word for the comment c_j, and β_{jt} is calculated as follows,

$$\mathbf{o}_{jt} = \tanh(\mathbf{W}_c \mathbf{h}_{jt} + \mathbf{b}_c)$$
$$\beta_{jt} = \frac{\exp(\mathbf{o}_{jt} \mathbf{o}_c^\mathsf{T})}{\sum_{k=1}^{Q_j} \exp(\mathbf{o}_k^j \mathbf{o}_c^\mathsf{T})}, \tag{4.34}$$

where $\mathbf{o}_{jt}$ is a hidden representation of $\mathbf{h}_{jt}$ obtained by feeding the hidden state $\mathbf{h}_{jt}$ to a fully embedding layer, and $\mathbf{u}_c$ is the weight parameter.

Sentence-Comment Co-attention   We observe that not all sentences in news contents are fake, and in fact, many sentences are true but only for supporting wrong claim sentences [40]. Thus, news sentences may not be equally important in determining and explaining whether a piece of news is fake or not. For example, the sentence "Michelle Obama is so vulgar she's not only being vocal.." is strongly related to the major fake claim "Pence: Michelle Obama Is The Most Vulgar First Lady We've Ever Had," while "The First Lady denounced the Republican presidential nominee" is a sentence that expresses some fact and is less helpful in detecting and explaining whether the news is fake.

Similarly, user comments may contain relevant information about the important aspects that explain why a piece of news is fake, while they may also be less informative and noisy. For example, a comment "Where did Pence say this? I saw him on CBS this morning and he didn't say these things.." is more explainable and useful to detect the fake news, than other comments such as "Pence is absolutely right."

Thus, we aim to select some news sentences and user comments that can explain why a piece of news is fake. As they provide a good explanation, they should also be helpful in detecting fake news. This suggests us to design attention mechanisms to give high weights of representations of news sentences and comments that are beneficial to fake news detection. Specifically, we use sentence-comment co-attention because it can capture the semantic affinity of sentences and comments and further help learn the attention weights of sentences and comments simultaneously.

We can construct the feature matrix of news sentences $\mathbf{S} = [\mathbf{s}_1; \cdots, \mathbf{s}_L] \in \mathbb{R}^{2d \times L}$ and the feature map of user comments $\mathbf{C} = \{\mathbf{c}_1, \cdots, \mathbf{c}_T\} \in \mathbb{R}^{2d \times T}$, the co-attention attends to the sentences and comments simultaneously. Similar to [83, 172], we first compute the affinity matrix $\mathbf{F} \in \mathbb{R}^{T \times L}$ as follows:

$$\mathbf{F} = \tanh(\mathbf{C}^\mathsf{T} \mathbf{W}_l \mathbf{S}), \tag{4.35}$$

where $\mathbf{W}_l \in \mathbb{R}^{2d \times 2d}$ is a weight matrix to be learned through the networks. Following the optimization strategy in [83], we can consider the affinity matrix as a feature and learn to predict sentence and comment attention maps as follows:

$$
\begin{aligned}
\mathbf{H}^s &= \tanh\left(\mathbf{W}_s\mathbf{S} + (\mathbf{W}_c\mathbf{C})\mathbf{F}\right) \\
\mathbf{H}^c &= \tanh\left(\mathbf{W}_c\mathbf{C} + (\mathbf{W}_s\mathbf{S})\mathbf{F}^\mathsf{T}\right),
\end{aligned} \tag{4.36}
$$

where $\mathbf{W}_s, \mathbf{W}_c \in \mathbb{R}^{k \times 2d}$ are the weight parameters. The attention weights of sentences and comments are calculated as follows:

$$
\begin{aligned}
\mathbf{a}^s &= softmax\left(\mathbf{w}_{hs}^\mathsf{T}\mathbf{H}^s\right) \\
\mathbf{a}^c &= softmax\left(\mathbf{w}_{hc}^\mathsf{T}\mathbf{H}^c\right),
\end{aligned} \tag{4.37}
$$

where $\mathbf{a}^s \in \mathbb{R}^{1 \times N}$ and $\mathbf{a}^c \in \mathbb{R}^{1 \times T}$ are the attention probabilities of each sentence s_i and comment c^j, respectively. $\mathbf{w}_{hs}, \mathbf{w}_{hc} \in \mathbb{R}^{1 \times k}$ are the weight parameters. The affinity matrix $\mathbf{F}$ transforms user comment attention space to news sentence attention space, and vice versa for $\mathbf{F}^\mathsf{T}$. Based on the above attention weights, the comment and sentence attention vectors are calculated as the weighted sum of the comment features and sentence features, i.e.,

$$
\hat{\mathbf{s}} = \sum_{i=1}^{L} \mathbf{a}_i^s \mathbf{s}_i, \qquad \hat{\mathbf{c}} = \sum_{j=1}^{T} \mathbf{a}_j^c \mathbf{c}_j, \tag{4.38}
$$

where $\hat{\mathbf{s}} \in \mathbb{R}^{1 \times 2d}$ and $\hat{\mathbf{c}} \in \mathbb{R}^{1 \times 2d}$ are the learned features for news sentences and user comments through co-attention.

Explainable Detection of Fake News   We have introduced how we can encode news contents by modeling the hierarchical structure from word level and sentence level, how we encode comments by word-level attention networks, and the component to model co-attention to learn sentences and comments representations. We further integrate these components together and predict fake news with the following objective:

$$
\hat{\mathbf{y}} = softmax([\hat{\mathbf{s}}, \hat{\mathbf{c}}]\mathbf{W}_f + \mathbf{b}_f), \tag{4.39}
$$

where $\hat{\mathbf{y}} = [\hat{\mathbf{y}}_0, \hat{\mathbf{y}}_1]$ is the predicted probability vector with $\hat{\mathbf{y}}_0$ and $\hat{\mathbf{y}}_1$ indicate the predicted probability of label being 0 (real news) and 1 (fake news) respectively. $y \in \{0, 1\}$ denotes the ground truth label of news. $[\hat{\mathbf{s}}, \hat{\mathbf{c}}]$ means the concatenation of learned features for news sentences and user comments. $\mathbf{b}_f \in \mathbb{R}^{1 \times 2}$ is the bias term. Thus, for each news piece, the goal is to minimize the cross-entropy loss function as follows:

$$
\mathcal{L}(\theta) = -y \log(\hat{\mathbf{y}}_1) - (1 - y)\log(1 - \hat{\mathbf{y}}_0), \tag{4.40}
$$

where θ denotes the parameters of the network.

APPENDIX A

Data Repository

The first and most important step to detect fake news is to collect a benchmark dataset. Despite several existing computational solutions on the detection of fake news, the lack of comprehensive and community-driven fake news datasets has become one of major roadblocks. In this appendix, we introduce a multi-dimensional data repository *FakeNewsNet*,[1] which contains two datasets with news content, social context, and spatiotemporal information [134]. For related datasets on fake news, rumors, etc., the readers can refer to several other survey papers such as [128, 183].

The constructed FakeNewsNet repository has the potential to boost the study of various open research problems related to fake news study. First, the rich set of features in the datasets provides an opportunity to experiment with different approaches for fake new detection, understand the diffusion of fake news in social network and intervene in it. Second, the temporal information enables the study of early fake news detection by generating synthetic user engagements from historical temporal user engagement patterns in the dataset [112]. Third, we can investigate the fake news diffusion process by identifying provenances, persuaders, and developing better fake news intervention strategies [131]. Our data repository can serve as a starting point for many exploratory studies for fake news, and provide a better, shared insight into disinformation tactics. This data repository is continuously updated with new sources and features. For a better comparison of the differences, we list existing popular fake news detection datasets below and compare them with the FakeNewsNet repository in Table A.1.

- ***BuzzFeedNews:***[2] This dataset comprises a complete sample of news published in Facebook from nine news agencies over a week close to the 2016 U.S. election from September 19–23 and September 26 and 27. Every post and the linked article were fact-checked claim-by-claim by five BuzzFeed journalists. It contains 1,627 articles, 826 mainstream, 356 left-wing, and 545 right-wing articles.

- ***LIAR:***[3] This dataset [163] is collected from fact-checking website PolitiFact. It has 12,800 human-labeled short statements collected from PolitiFact and the statements are labeled into six categories ranging from completely false to completely true as pants on fire, false, barely-true, half-true, mostly true, and true.

---

[1] https://github.com/KaiDMML/FakeNewsNet
[2] https://github.com/BuzzFeedNews/2016-10-facebook-fact-check/tree/master/data
[3] https://www.cs.ucsb.edu/~william/software.html

- **BS Detector:**[4] This dataset is collected from a browser extension called BS detector developed for checking news veracity. It searches all links on a given web page for references to unreliable sources by checking against a manually compiled list of domains. The labels are the outputs of the BS detector, rather than human annotators.

- **CREDBANK:**[5] This is a large-scale crowd-sourced dataset [91] of around 60 million tweets that cover 96 days starting from October 2015. The tweets are related to over 1,000 news events. Each event is assessed for credibilities by 30 annotators from Amazon Mechanical Turk.

- **BuzzFace:**[6] This dataset [120] is collected by extending the BuzzFeed dataset with comments related to news articles on Facebook. The dataset contains 2263 news articles and 1.6 million comments discussing news content.

- **FacebookHoax:**[7] This dataset [147] comprises information related to posts from the Facebook pages related to scientific news (non-hoax) and conspiracy pages (hoax) collected using Facebook Graph API. The dataset contains 15,500 posts from 32 pages (14 conspiracy and 18 scientific) with more than 2,300,000 likes.

- **NELA-GT-2018:**[8] This dataset collects articles between February 2018 through November 2018 from 194 news and media outlets including mainstream, hyper-partisan, and conspiracy sources, resulting in 713 k articles. The ground truth labels are integrated from eight independent assessments.

From Table A.1, we observe that no existing public dataset can provide all possible features of news content, social context, and spatiotemporal information. Existing datasets have some limitations that we try to address in our data repository. For example, BuzzFeedNews only contains headlines and text for each news piece and covers news articles from very few news agencies. LIAR dataset contains mostly short statements instead of entire news articles with the meta attributes. BS Detector data is collected and annotated by using a developed news veracity checking tool, rather than using human expert annotators. CREDBANK dataset was originally collected for evaluating tweet credibilities and the tweets in the dataset are not related to the fake news articles and hence cannot be effectively used for fake news detection. BuzzFace dataset has basic news contents and social context information but it does not capture the temporal information. The FacebookHoax dataset consists very few instances about the conspiracy theories and scientific news.

To address the disadvantages of existing fake news detection datasets, the proposed FakeNewsNet repository collects multi-dimension information from news content, social context,

---

[4]https://github.com/bs-detector/bs-detector
[5]http://compsocial.github.io/CREDBANK-data/
[6]https://github.com/gsantia/BuzzFace
[7]https://github.com/gabll/some-like-it-hoax
[8]https://dataverse.harvard.edu/dataset.xhtml?persistentId=doi:10.7910/DVN/ULHLCB

Table A.1: The comparison with representative fake news detection datasets

Datasets	News Content		Social Context				Spatiotemporal	
	Linguistic	Visual	User	Post	Response	Network	Spatial	Temporal
BuzzFeedNews	✓							
LIAR	✓							
BS Detector	✓							
CREDBANK	✓		✓	✓			✓	✓
BuzzFace	✓			✓	✓			✓
FacebookHoax	✓		✓	✓	✓			
NELA-GT-2018	✓							
FakeNewsNet	✓	✓	✓	✓	✓	✓	✓	✓

and spatiotemporal information from different types of news domains such as political and entertainment sources.

Data Integration   In this part, we introduce the dataset integration process for the FakeNewsNet repository. We demonstrate in Figure A.1 how we can collect news contents with reliable ground truth labels, and how we obtain additional social context and spatialtemporal information.

News Content: to collect reliable ground truth labels for fake news, we utilize fact-checking websites to obtain news contents for fake news and true news such as *PolitiFact*[9] and *GossipCop*.[10]

In PolitiFact, journalists and domain experts review the political news and provide fact-checking evaluation results to claim news articles as fake[11] or real.[12] We utilize these claims as ground truths for fake and real news pieces. In PolitiFact's fact-checking evaluation result, the source URLs of the web page that published the news articles are provided, which can be used to fetch the news contents related to the news articles. In some cases, the web pages of source news articles are removed and are no longer available. To tackle this problem, we (i) check if the removed page was archived and automatically retrieve content at the Wayback Machine;[13] and (ii) make use of Google web search in automated fashion to identify news article that is most related to the actual news.

---

[9]https://www.politifact.com/
[10]https://www.gossipcop.com/
[11]Available at https://www.politifact.com/subjects/fake-news/.
[12]Available at https://www.politifact.com/truth-o-meter/rulings/true/.
[13]https://archive.org/web/

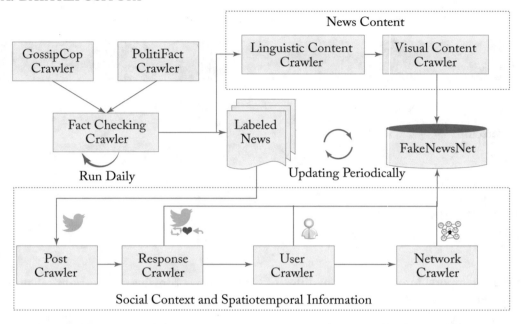

Figure A.1: The flowchart of data integration process for FakeNewsNet. It mainly describes the collection of news content, social context and spatiotemporal information.

GossipCop is a website for fact-checking entertainment stories aggregated from various media outlets. GossipCop provides rating scores on the scale of 0–10 to classify a news story as the degree from fake to real. In order to collect true entertainment news pieces, we crawl the news articles from E! Online,[14] which is a well-known trusted media website for publishing entertainment news pieces. We consider all the articles from E! Online as real news sources. We collect all the news stories from GossipCop with rating scores less than 5 as the fake news stories. Since GossipCop does not explicitly provide the URL of the source news article, so we search the news headline in Google or the Wayback Machine archive to obtain the news source information.

Social Context: the user engagements related to the fake and real news pieces from fact-checking websites are collected using search API provided by social media platforms such as the Twitter's Advanced Search API.[15] The search queries for collecting user engagements are formed from the headlines of news articles, with special characters removed from the search query to filter out the noise. After we obtain the social media posts that directly spread news pieces, we further fetch the user *response* toward these posts such as replies, likes, and reposts.

[14]https://www.eonline.com/
[15]https://twitter.com/search-advanced?lang=en

In addition, when we obtain all the users engaging in news dissemination process, we collect all the metadata for user profiles, user posts, and the social network information.

Spatiotemporal Information: the spatiotemporal information includes spatial and temporal information. For spatial information, we obtain the locations explicitly provided in user profiles. The temporal information indicates that we record the timestamps of user engagements, which can be used to study how fake news pieces propagate on social media, and how the topics of fake news are changing over time. Since fact-checking websites periodically update newly coming news articles, so we dynamically collect these newly added news pieces and update the FakeNewsNet repository as well. In addition, we keep collecting the user engagements for all the news pieces periodically in the FakeNewsNet repository such as the recent social media posts, and second order user behaviors such as replies, likes, and retweets. For example, we run the news content crawler and update Tweet collector per day. The spatiotemporal information provides useful and comprehensive information for studying fake news problem from a temporal perspective.

APPENDIX B

Tools

In this appendix, we introduce some representative online tools for tracking and detecting fake news on social media.

Hoaxy[1]   Hoaxy[1] aims to build a uniform and extensible platform to collect and track misinformation and fact-checking [125], with visualization techniques to understand the misinformation propagation on social media.

Data Scraping. The major components included a tracker for the Twitter API, and a set of crawlers for both fake news and fact checking websites and databases (see Figure B.1). The system collects data from two main sources: news websites and social media. From the first group we can obtain data about the origin and evolution of both fake news stories and their fact checking. From the second group we collect instances of these news stories (i.e., URLs) that are being shared online. To collect data from such disparate sources, different technologies are used: web scraping, web syndication, and, where available, APIs of social networking platforms. To collect data on news stories they use rich site summary (RSS), which allows a unified protocol instead of manually adapting our scraper to the multitude of web authoring systems used on the Web. RSS feeds contain information about updates made to news stories. The data is collected from news sites using the following two steps: when a new website is added to our list of monitored sources, a "deep" crawl of its link structure is performed using a custom Python spider written with the Scrapy framework; at this stage, the URL of the RSS feed is identified if it is available. Once all existing stories have been acquired, a "light" crawl is performed every two hour by checking its RSS feed only. To perform the "deep" crawl, we use a depth first strategy. The "light" crawl is instead performed using a breadth-first approach.

Analysis Dashboard. Hoaxy provides various visualization interfaces to demonstrate the news spreading process. As shown in Figure B.2, we demonstrate the major functionalities on the analysis dashboard. On the top, users can search any news articles by providing specific keywords. On the left side, it demonstrates the temporal trendiness of the user engagements for the news articles. On the right side, it illustrates the propagation network on Twitter, which clearly convey the information on who spreads the news tweets from whom. In addition, they also evaluate the bot score for all users with BotMeter [31].

---

[1]https://hoaxy.iuni.iu.edu/

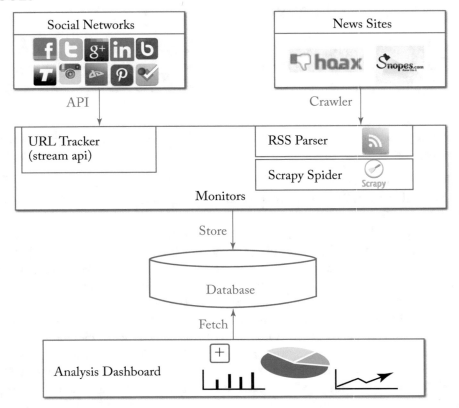

Figure B.1: The framework of Hoaxy. Based on [125].

FakeNewsTracker   FakeNewsTracker[2] is a system for fake news data collection, detection, and visualization on social media [133]. It mainly consists of the following components (see Figure B.3): (1) fake news collection; (2) fake news detection; and (3) fake news visualization.

Collecting Fake News Data. Fake news is widely spread across various online platforms. We use some of the fact-checking websites like PolitiFact as a source for collecting fake news information. On these fact-checking sites, fake news information is provided by the trusted authors and relevant claims are made by the authors on why the mentioned news is not true. The detailed collection procedure is described in Figure A.1.

Detecting Fake News. A deep learning model is proposed to learn neural textual features from news content, and temporal representations from social context simultaneously to predict fake news. An auto-encoder [67] is used to learn the feature representation of news articles, by reconstructing the news content, and LSTM is utilized to learn the temporal features of user

---

[2]http://blogtrackers.fulton.asu.edu:3000/

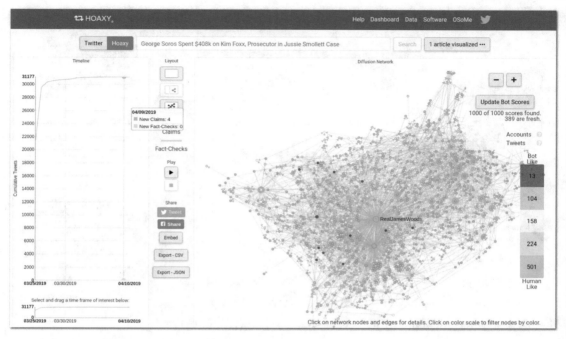

Figure B.2: The main dashboard of Hoaxy website.

engagements. Finally, the learned feature representations of news and social engagements are fused to predict fake news.

Visualization Fake News in Twitter. We have developed a fake news visualization as shown in Figure B.4 for the developing insights on the collected data through various interfaces. We demonstrate the temporal trends of the number of tweets spreading fake and real news in a specific time period, as in Figure B.4a.

In addition, we can explore the social network structure among users in the propagation network (see Figure B.4b for an example), and further compare the differences between the users who interact with the fake news and the true news.

For identifying the differences in the news content of the true news and the fake news we have used word cloud representation of the words for the textual data. We search for fake news within a time frame and identify the relevant data. In addition, we provide the comparison of feature significance and model performance as part of this dashboard. Moreover, we could see how fake news is spread around certain areas using the geo-locations of tweets.

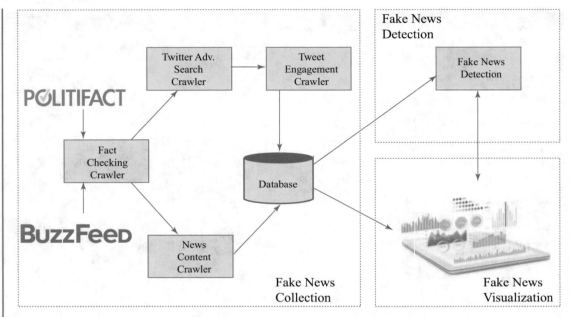

Figure B.3: The framework of FakeNewsTracker. Based on [133].

dEFEND   dEFEND[3] is a fake news detection system that are also able to provide explainable user comments on Twitter. dEFEND (see Figure B.5) mainly consists of two major components: a web-based user interface and a backend which integrates our fake news detection model.

The web-based interface provides users with explainable fact-checking of news. A user can input either the tweet URL or the title of the news. A screenshot was shown in Figure B.6. On typical fact-checking websites, a user just sees the check-worthy score of news (like Gossip Cop[4]) or each sentence (like ClaimBuster.[5]) In our approach, the user cannot only see the detection result (in the right of Figure B.6a), but also can find all the arguments that support the detection result, including crucial sentences in the article (in the middle of Figure B.6b) and explainable comments from social media platforms (in the right of Figure B.6b). At last, the user can also review the results and find related news/claims.

The system also provides exploratory search functions including news propagation network, trending news, top claims and related news. The news propagation network (in the left of Figure B.6b) is to help readers understand the dynamics of real and fake news sharing, as fake news are normally dominated by very active users, while real news/fact checking is a more

---

[3]http://fooweb-env.qnmbmwmxj3.us-east-2.elasticbeanstalk.com/
[4]https://www.gossipcop.com/
[5]https://idir-server2.uta.edu/claimbuster/

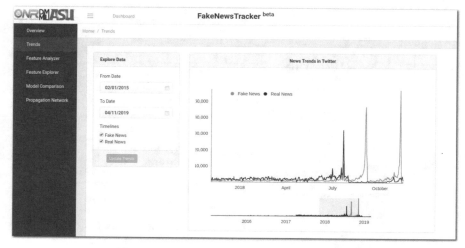

(a) User interface of trend on news spreading

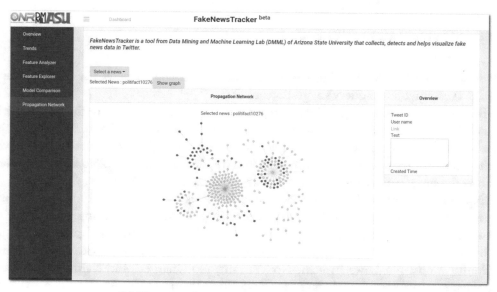

(b) User interface on news propagation networks

Figure B.4: Demonstration of FakeNewsTracker system.

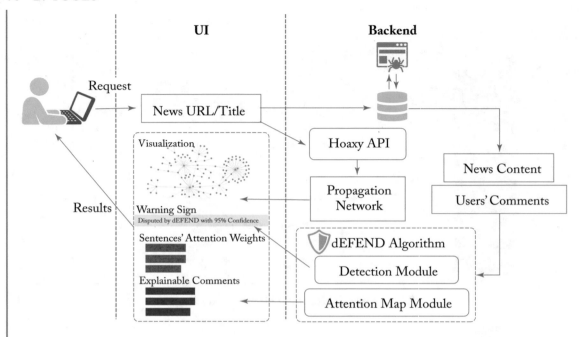

Figure B.5: The framework of dEFEND.

grass-roots activity [125]. Trending news, top claims, and related news (in the lower left of Figure B.6a) can give some query suggestions to users.

The backend consists of multiple components: (1) a database to store the pre-trained results as well as a crawler to extract unseen news and its comments; (2) the dEFEND algorithm module based on explainable deep learning fake news detection (details in Section 4.3.2), which gives the detection result and explanations simultaneously; and (3) an exploratory component that shows the propagation network of the news, trending and related news.

Exploratory Search. The system also provides users with browsing functions. Consider a user who doesn't know what to check specifically. By browsing the trending news, top claims and news related to the previous search right below the input box, the user can get some ideas about what he could do. News can be the coverage of an event, such as "Seattle Police Begin Gun Confiscations: No Laws Broken, No Warrant, No Charges" and claim is the coverage around what a celebrity said, such as "Actor Brad Pitt: 'Trump Is Not My President, We Have No Future With This....'" Users can search these titles by clicking on them. The news related to the user's previous search is recommended. For example, news "Obama's Health Care Speech to Congress" is related to the query "It's Better For Our Budget If Cancer Patients Die More Quickly."

(a) User interface of search: the input box (upper left), query suggestions (lower left), and intuitive propagation network (right).

(b) Explainable Fact Checking: news content (left), explainable sentences (upper right), and comments (lower right).

Figure B.6: Demonstration of dEFEND system.

Explainable Fact-Checking. Consider a user who wants to check whether Tom Price has said "It's Better For Our Budget If Cancer Patients Die More Quickly." The user first enters the tweet URL or the title of a news in the input box in Figure B.6a. The system would return the check-worthy score, the propagation network, sentences with explainable scores, and comments with explainable scores to the user in Figure B.6b. The user can zoom in the network to check the details of the diffusion path. Each sentence is shown in the table along with its score. The higher the score, the more likely the sentence contains check-worthy factual claims. The lower the score, the more non-factual and subjective the sentence is. The user can sort the sentences either by the order of appearance or by the score. Comments' explainable scores are similar to sentences'. The top-5 comments are shown in the descending order of their explainable score.

NewsVerify   NewsVerify[6] is a real-time news verification system which can detect the credibility of an event by providing some keywords about it [184].

NewsVerify mainly contains three stages: (1) crawling data; (2) building an ensemble model; and (3) visualizing the results. Given the keywords and time range of a news event, the related microblogs can be collected through the search engine of Sina Weibo. Based on these messages, the key users and microblogs can be extracted for further analysis. The key users are used for information source certification while the key microblogs are used for propagation and content certification. All the data above are crawled through distributed data acquisition system which will be illustrated below. After three individual models have been developed, the scores from the above mentioned models are combined via weighted combination. Finally, an event level credibility score is provided, and each single model will also have a credibility score that measure the credibility of corresponding aspect. To improve the user experience of our application, the results are visualized from various perspectives, which provide useful information of events for further investigation.

Data Acquisition. Three kinds of information are collected: microblogs, propagation, and microbloggers. Like most distributed system, NewsVerify also has master node and child nodes. The master node is responsible for task distribution and results integration while child node process the specific task and store the collected data in the appointed temporary storage space. The child node will inform the master node after all tasks finished. Then, master node will merge all slices of data from temporary storage space and stored the combined data in permanent storage space. After above operations, the temporary storage will be deleted. The distributed system is based on ZooKeeper,[7] a centralized service for maintaining configuration information, naming, providing distributed synchronization, and providing group services. As the attributes of frequent data interaction, stored, read, we adopt efficient key-val database Redis to handle the real-time data acquisition task. Redis, working with an in-memory dataset, can achieve outstanding performance.

Model Ensemble. Different individual models are built to verify the truthfulness of news pieces from the perspective of news content, news propagation, and information source (see Figure B.7). The *content-based* model is based on hierarchical propagation networks [58]. The credibility network has three layers: message layer, sub-event layer and event layer. Following that, the semantic and structure features are exploited to adjust the weights of links in the network. Given a news event and its related microblogs, sub-events are generated by clustering algorithm. Sub-event layer is constructed to capture implicit semantic information within an event. Four types of network links are made to reflect the relation between network nodes. The intra-level links (Message to Message, Sub-event to Sub-event) reflect the relations among entities of a same type while the inter level links (Message to Sub-event, Sub-event to Event) reflect the impact from level to level. After the network constructed, all entities are initialize with cred-

---

[6]https://www.newsverify.com/
[7]http://zookeeper.apache.org/

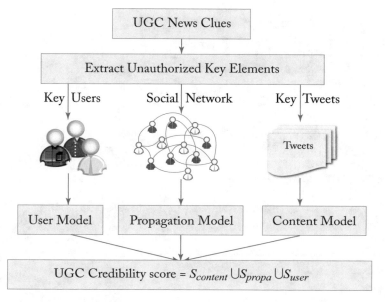

Figure B.7: The framework of NewsVerify system. Based on [184].

ibility values using classification results. We formulate this propagation as a graph optimization problem and provides a global optimal solution to it. The *propagation-based* model propose to compute a propagation influence score over time to capture the temporal trends. The *information source based* model utilize the sentiment and activeness degree as features to help predict fake news. From the aforementioned models, an individual score is obtained. Then a weighted logistic regression model can be used to ensemble the result and produce an overall score for the news piece.

Interface Visualization. Figure B.8 illustrate the interface of NewsVerify system. It allows users to report fake news, and search specific news to verify by providing keywords to the system. It also automatically show the degree of veracity for Weibo data of different categories. For each Weibo in the time-line, NewsVerify shows the credibility score to justify how likely the Weibo is related to fake news. In addition, it allows interested users to click "View detailed analysis" to learn more about the news. As shown in Figure B.9, it mainly demonstrates: (1) the introduction of the news event including the time, source, related news, etc.; (2) the changes of trends and topics over time related to the news events; (3) the profiles and aggregated statistics of users who are engaged in the news spreading process such as the key communicator, sex ratio, certification ratio; and (4) the images or videos related to the news events.

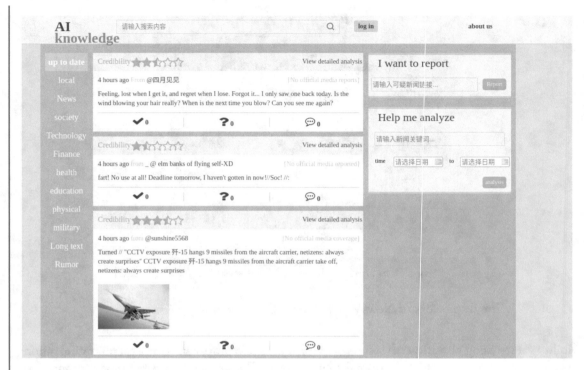

Figure B.8: The interface of NewsVerify system.

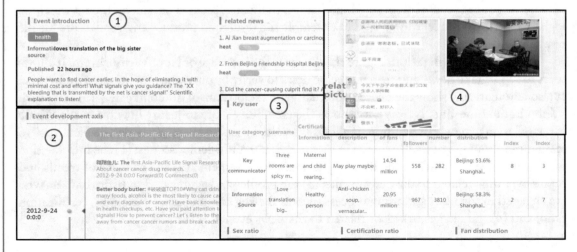

Figure B.9: Demonstration of detail news analysis of NewsVerify system.

APPENDIX C

Relevant Activities

Fake news has been attracting interests from experts and practitioners of multiple disciplines. Relevant activities are organized to advance the detection and mitigation of fake news. Specifically, we introduce these efforts in three general categories: *Educational Programs*, *Computational Competitions*, and *Research Workshops and Tutorials*.

Educational Programs   These programs aim to help design and train interested people about how to identify fake news. The educational programs include handbooks [56, 94], interactive games. For example, in [56], the researchers from UNESCO[1] build a series of curricula and handbooks for journalism education and training. Similarly, a cookbook is built to help identify fake news from the perspectives of transparency, engagement, education, and tools.[2] Interactive online programs are designed that encode some heuristic features of fake news detection to help learn common tricks of identifying fake news via game playing. "Bad News"[3] is an online game that allows users to act as a fake news creator to build a fake "credibility" step by step.

Computational Competitions   To encourage researchers or students to build computational algorithms to address fake news problems, several competitions are organized online or in conjunction with some conferences. For example, the fake news challenge[4] aims to detect the stance of pairs of headline and body text, which attracts many researchers to create effective solutions to improve the performance [115, 124]. As another example, Bytedance organized a fake news classification challenge in conjunction with the ACM WSDM conference,[5] aiming to identify if a given news pieces is related to another piece of fake news. Moreover, the SBP competition on disinformation is regularly held to encourage researchers to combat fake news.[6]

Research Workshops and Tutorials   To bring researchers and practitioners together to brainstorm novel ways of dealing with fake news, different research workshops and tutorials are held from various perspectives. One of the earliest and influential workshop [88] aims to define the foundations, actions, and research directions on combating fake news. The social cyber-security working group[7] has brought together experts to deal with various cyber security threats on so-

---

[1]The United Nations Educational, Scientific and Cultural Organization
[2]https://newscollab.org/best-practices/
[3]https://getbadnews.com/
[4]http://www.fakenewschallenge.org/
[5]https://www.kaggle.com/c/fake-news-pair-classification-challenge/
[6]http://sbp-brims.org/2019/challenge/challenge2_Disinformation.html
[7]https://sites.google.com/view/social-cybersec/

cial media including disinformation and fake news. Recently, the National Academy of Sciences Colloquia hosted a panel on the communication and science of misinformation and fake news.[8] Several tutorials are offered in conjunction with top-tier conferences such as ACM KDD 2019,[9] ACM WSDM 2019 [187], AAAI 2018,[10] and IEEE ICDM 2017.[11] For example, the tutorials of KDD 2019 and WSDM 2019 focus on the fundamental theories, detection strategies, and open issues, the AAAI 2018 tutorial discusses fake news from artificial intelligence and database perspectives, and the ICDM 2017 tutorial presents the detection and spreading patterns of misinformation on social media.

---

[8]http://www.cvent.com/events/advancing-the-science-and-practice-of-science-communication-misinformation-about-science-in-the-publ/event-summary-c4d9df4d8baf4567ab82042e4f4efb78.aspx

[9]https://www.fake-news-tutorial.com/

[10]https://john.cs.olemiss.edu/~nhassan/file/aaai2018tutorial.html

[11]http://www.public.asu.edu/~liangwu1/ICDM17MisinformationTutorial.html

Bibliography

[1] Mohammad Ali Abbasi and Huan Liu. Measuring user credibility in social media. In *SBP*, pages 441–448, Springer, 2013. DOI: 10.1007/978-3-642-37210-0_48 47

[2] Sadia Afroz, Michael Brennan, and Rachel Greenstadt. Detecting hoaxes, frauds, and deception in writing style online. In *ISSP*, 2012. DOI: 10.1109/sp.2012.34 17

[3] Anderson, Jonathan. Lix and rix: Variations on a little-known readability index, *Journal of Reading*, 26(6), pages 490–496, JSTOR, 1983. 18

[4] Ameeta Agrawal, Aijun An, and Manos Papagelis. Learning emotion-enriched word representations. In *Proc. of the 27th International Conference on Computational Linguistics*, pages 950–961, 2018. 36

[5] Hadeer Ahmed, Issa Traore, and Sherif Saad. Detection of online fake news using n-gram analysis and machine learning techniques. In *International Conference on Intelligent, Secure, and Dependable Systems in Distributed and Cloud Environments*, pages 127–138, Springer, 2017. DOI: 10.1007/978-3-319-69155-8_9 9

[6] Hunt Allcott and Matthew Gentzkow. Social media and fake news in the 2016 election. *Technical Report*, National Bureau of Economic Research, 2017. DOI: 10.3386/w23089 2

[7] Solomon E. Asch and H. Guetzkow. Effects of group pressure upon the modification and distortion of judgments. *Groups, Leadership, and Men*, pages 222–236, 1951. 3

[8] Dzmitry Bahdanau, Kyunghyun Cho, and Yoshua Bengio. Neural machine translation by jointly learning to align and translate. *ArXiv Preprint ArXiv:1409.0473*, 2014. 12, 74

[9] Meital Balmas. When fake news becomes real: Combined exposure to multiple news sources and political attitudes of inefficacy, alienation, and cynicism. *Communication Research*, 41(3):430–454, 2014. DOI: 10.1177/0093650212453600 2

[10] Michele Banko, Michael J. Cafarella, Stephen Soderland, Matthew Broadhead, and Oren Etzioni. Open information extraction from the Web. In *IJCAI*, 2007. DOI: 10.1145/1409360.1409378 21

[11] Herbert Bay, Tinne Tuytelaars, and Luc Van Gool. Surf: Speeded up robust features. In *European Conference on Computer Vision*, pages 404–417, Springer, 2006. DOI: 10.1007/11744023_32 15

[12] Kristy Beers Fägersten. Who's swearing now?: The social aspects of conversational swearing, 2012. 28

[13] Omar Benjelloun, Hector Garcia-Molina, David Menestrina, Qi Su, Steven Euijong Whang, and Jennifer Widom. Swoosh: A generic approach to entity resolution. *VLDB*, 2009. DOI: 10.1007/s00778-008-0098-x 23

[14] Alessandro Bessi and Emilio Ferrara. Social bots distort the 2016 us presidential election online discussion. *First Monday*, 21(11), 2016. DOI: 10.5210/fm.v21i11.7090 4

[15] Mikhail Bilenko and Raymond J. Mooney. Adaptive duplicate detection using learnable string similarity measures. In *KDD*, 2003. DOI: 10.1145/956750.956759 23

[16] Prakhar Biyani, Kostas Tsioutsiouliklis, and John Blackmer. 8 amazing secrets for getting more clicks: Detecting clickbaits in news streams using article informality. In *AAAI*, 2016. 17

[17] Kurt Bollacker, Colin Evans, Praveen Paritosh, Tim Sturge, and Jamie Taylor. Freebase: A collaboratively created graph database for structuring human knowledge. In *Proc. of the ACM SIGMOD International Conference on Management of Data*, pages 1247–1250, 2008. DOI: 10.1145/1376616.1376746 23

[18] Paul R. Brewer, Dannagal Goldthwaite Young, and Michelle Morreale. The impact of real news about "fake news": Intertextual processes and political satire. *International Journal of Public Opinion Research*, 25(3):323–343, 2013. DOI: 10.1093/ijpor/edt015 2

[19] David Guy Brizan and Abdullah Uz Tansel. A survey of entity resolution and record linkage methodologies. *Communications of the IIMA*, 2015. 22

[20] Michael M. Bronstein, Joan Bruna, Yann LeCun, Arthur Szlam, and Pierre Vandergheynst. Geometric deep learning: Going beyond Euclidean data. *IEEE Signal Processing Magazine*, 34(4):18–42, 2017. DOI: 10.1109/msp.2017.2693418 48

[21] Juan Cao, Junbo Guo, Xirong Li, Zhiwei Jin, Han Guo, and Jintao Li. Automatic rumor detection on microblogs: A survey. *ArXiv Preprint ArXiv:1807.03505*, 2018. 14

[22] Carlos Castillo, Marcelo Mendoza, and Barbara Poblete. Information credibility on twitter. In *WWW*, 2011. DOI: 10.1145/1963405.1963500 25, 37

[23] Fabio Celli and Massimo Poesio. Pr2: A language independent unsupervised tool for personality recognition from text. *ArXiv Preprint ArXiv:1402.2796*, 2014. 26

[24] Huimin Chen, Maosong Sun, Cunchao Tu, Yankai Lin, and Zhiyuan Liu. Neural sentiment classification with user and product attention. In *EMNLP*, 2016. DOI: 10.18653/v1/d16-1171 74

[25] Yimin Chen, Niall J. Conroy, and Victoria L. Rubin. Misleading online content: Recognizing clickbait as false news. In *Proc. of the ACM on Workshop on Multimodal Deception Detection*, pages 15–19, 2015. DOI: 10.1145/2823465.2823467 17

[26] Justin Cheng, Michael Bernstein, Cristian Danescu-Niculescu-Mizil, and Jure Leskovec. Anyone can become a troll: Causes of trolling behavior in online discussions. In *CSCW*, 2017. DOI: 10.1145/2998181.2998213 5

[27] Kyunghyun Cho, Bart Van Merriënboer, Caglar Gulcehre, Dzmitry Bahdanau, Fethi Bougares, Holger Schwenk, and Yoshua Bengio. Learning phrase representations using RNN encoder-decoder for statistical machine translation. *ArXiv Preprint ArXiv:1406.1078*, 2014. DOI: 10.3115/v1/d14-1179 74

[28] Zi Chu, Steven Gianvecchio, Haining Wang, and Sushil Jajodia. Detecting automation of twitter accounts: Are you a human, bot, or cyborg? *IEEE Transactions on Dependable and Secure Computing*, 9(6):811–824, 2012. DOI: 10.1109/tdsc.2012.75 5

[29] Giovanni Luca Ciampaglia, Prashant Shiralkar, Luis M. Rocha, Johan Bollen, Filippo Menczer, and Alessandro Flammini. Computational fact checking from knowledge networks. *PloS One*, 10(6):e0128193, 2015. DOI: 10.1371/journal.pone.0141938 22, 72

[30] Niall J. Conroy, Victoria L. Rubin, and Yimin Chen. Automatic deception detection: Methods for finding fake news. *Proc. of the Association for Information Science and Technology*, 52(1):1–4, 2015. DOI: 10.1002/pra2.2015.145052010082 2

[31] Clayton Allen Davis, Onur Varol, Emilio Ferrara, Alessandro Flammini, and Filippo Menczer. Botornot: A system to evaluate social bots. In *Proc. of the 25th International Conference Companion on World Wide Web*, pages 273–274, International World Wide Web Conferences Steering Committee, 2016. DOI: 10.1145/2872518.2889302 85

[32] Michela Del Vicario, Alessandro Bessi, Fabiana Zollo, Fabio Petroni, Antonio Scala, Guido Caldarelli, H. Eugene Stanley, and Walter Quattrociocchi. The spreading of misinformation online. *Proc. of the National Academy of Sciences*, 113(3):554–559, 2016. DOI: 10.1073/pnas.1517441113 5

[33] Michela Del Vicario, Gianna Vivaldo, Alessandro Bessi, Fabiana Zollo, Antonio Scala, Guido Caldarelli, and Walter Quattrociocchi. Echo chambers: Emotional contagion and group polarization on facebook. *Scientific Reports*, 6, 2016. DOI: 10.1038/srep37825 5

[34] Nikos Deligiannis, Tien Huu Do, Duc Minh Nguyen, and Xiao Luo. Deep learning for geolocating social media users and detecting fake news. https://www.sto.nato.int/p ublications/.../STO-MP-IST-160/MP-IST-160-S3-5.pdf 48

[35] Thomas Deselaers, Tobias Gass, Philippe Dreuw, and Hermann Ney. Jointly optimising relevance and diversity in image retrieval. In *Proc. of the ACM International Conference on Image and Video Retrieval*, page 39, 2009. DOI: 10.1145/1646396.1646443 16

[36] Xin Luna Dong, Evgeniy Gabrilovich, Geremy Heitz, Wilko Horn, Kevin Murphy, Shaohua Sun, and Wei Zhang. From data fusion to knowledge fusion. *Proc. of the VLDB Endowment*, 7(10):881–892, 2014. DOI: 10.14778/2732951.2732962 23

[37] Mohamed G. Elfeky, Vassilios S. Verykios, and Ahmed K. Elmagarmid. Tailor: A record linkage toolbox. In *ICDE*, 2002. DOI: 10.1109/icde.2002.994694 23

[38] Robert M. Entman. Framing: Toward clarification of a fractured paradigm. *Journal of Communication*, 43(4):51–58, 1993. DOI: 10.1111/j.1460-2466.1993.tb01304.x 3

[39] Ivan P. Fellegi and Alan B. Sunter. A theory for record linkage. *Journal of the American Statistical Association*, 1969. DOI: 10.2307/2286061 23

[40] Song Feng, Ritwik Banerjee, and Yejin Choi. Syntactic stylometry for deception detection. In *ACL*, pages 171–175, Association for Computational Linguistics, 2012. 17, 76

[41] Emilio Ferrara, Onur Varol, Clayton Davis, Filippo Menczer, and Alessandro Flammini. The rise of social bots. *Communications of the ACM*, 59(7):96–104, 2016. DOI: 10.1145/2818717 4

[42] Emilio Ferrara and Zeyao Yang. Quantifying the effect of sentiment on information diffusion in social media. *PeerJ Computer Science*, 1:e26, 2015. DOI: 10.7717/peerj-cs.26 35

[43] Johannes Fürnkranz. A study using n-gram features for text categorization. *Austrian Research Institute for Artificial Intelligence*, 3(1998):1–10, 1998. 9

[44] Matthew Gentzkow, Jesse M. Shapiro, and Daniel F. Stone. Media bias in the marketplace: Theory. *Technical Report*, National Bureau of Economic Research, 2014. DOI: 10.3386/w19880 4, 27

[45] Clayton J. Hutto and Eric Gilbert. Vader: A parsimonious rule-based model for sentiment analysis of social media text. *8th International AAAI Conference on Weblogs and Social Media*, 2014. 49, 53

[46] Gisel Bastidas Guacho, Sara Abdali, Neil Shah, and Evangelos E. Papalexakis. Semi-supervised content-based detection of misinformation via tensor embeddings. In *IEEE/ACM International Conference on Advances in Social Networks Analysis and Mining (ASONAM)*, pages 322–325, 2018. DOI: 10.1109/asonam.2018.8508241 10

[47] Chuan Guo, Juan Cao, Xueyao Zhang, Kai Shu, and Miao Yu. Exploiting emotions for fake news detection on social media. *ArXiv Preprint ArXiv:1903.01728*, 2019. 35

[48] Han Guo, Juan Cao, Yazi Zhang, Junbo Guo, and Jintao Li. Rumor detection with hierarchical social attention network. In *Proc. of the 27th ACM International Conference on Information and Knowledge Management*, pages 943–951, 2018. DOI: 10.1145/3269206.3271709 72

[49] Shashank Gupta, Raghuveer Thirukovalluru, Manjira Sinha, and Sandya Mannarswamy. Cimtdetect: A community infused matrix-tensor coupled factorization based method for fake news detection. In *IEEE/ACM International Conference on Advances in Social Networks Analysis and Mining (ASONAM)*, pages 278–281, 2018. DOI: 10.1109/asonam.2018.8508408 43, 44

[50] Kaiming He, Xiangyu Zhang, Shaoqing Ren, and Jian Sun. Deep residual learning for image recognition. In *Proc. of the IEEE Conference on Computer Vision and Pattern Recognition*, pages 770–778, 2016. DOI: 10.1109/cvpr.2016.90 16

[51] Heylighen, Francis and Dewaele, Jean-Marc. Formality of language: Definition, measurement and behavioral determinants, *Interner Bericht, Center "Leo Apostel," Vrije Universiteit Brüssel*, Citeseer, 1999. 18

[52] Johannes Hoffart, Fabian M. Suchanek, Klaus Berberich, and Gerhard Weikum. Yago2: A spatially and temporally enhanced knowledge base from Wikipedia. *Artificial Intelligence*, 194:28–61, 2013. DOI: 10.1016/j.artint.2012.06.001 23

[53] Seyedmehdi Hosseinimotlagh and Evangelos E. Papalexakis. Unsupervised contentbased identification of fake news articles with tensor decomposition ensembles. 2018. http://snap.stanford.edu/mis2/files/MIS2_paper_2.pdf 10, 63, 64

[54] Xia Hu, Jiliang Tang, Huiji Gao, and Huan Liu. Unsupervised sentiment analysis with emotional signals. In *Proc. of the 22nd WWW*, pages 607–618, International World Wide Web Conferences Steering Committee, 2013. DOI: 10.1145/2488388.2488442 37

[55] Clayton J. Hutto and Eric Gilbert. Vader: A parsimonious rule-based model for sentiment analysis of social media text. In *8th International AAAI Conference on Weblogs and Social Media*, 2014. 36, 67

[56] Cherilyn Ireton and Julie Posetti. Journalism, "fake news" and disinformation, 2018. https://cdn.isna.ir/d/2019/01/19/0/57816097.pdf 95

[57] Zhiwei Jin, Juan Cao, Han Guo, Yongdong Zhang, and Jiebo Luo. Multimodal fusion with recurrent neural networks for rumor detection on microblogs. In *Proc. of the 25th ACM International Conference on Multimedia*, pages 795–816, 2017. DOI: 10.1145/3123266.3123454 16

[58] Zhiwei Jin, Juan Cao, Yu-Gang Jiang, and Yongdong Zhang. News credibility evaluation on microblog with a hierarchical propagation model. In *ICDM*, pages 230–239, IEEE, 2014. DOI: 10.1109/icdm.2014.91 92

[59] Zhiwei Jin, Juan Cao, Yongdong Zhang, and Jiebo Luo. News verification by exploiting conflicting social viewpoints in microblogs. In *AAAI*, pages 2972–2978, 2016. 2, 25, 39, 40, 49, 72

[60] Zhiwei Jin, Juan Cao, Yongdong Zhang, Jianshe Zhou, and Qi Tian. Novel visual and statistical image features for microblogs news verification. *IEEE Transactions on Multimedia*, 19(3):598–608, 2017. DOI: 10.1109/tmm.2016.2617078 14

[61] Daniel Kahneman and Amos Tversky. Prospect theory: An analysis of decision under risk. *Econometrica: Journal of the Econometric Society*, pages 263–291, 1979. 3

[62] Jaap Kamps, Maarten Marx, Robert J. Mokken, and Maarten de Rijke. Using wordnet to measure semantic orientations of adjectives. In *Proc. of the 4th International Conference on Language Resources and Evaluation, (LREC)*, Lisbon, Portugal, May 26–28, 2004. 36

[63] Jean-Noel Kapferer. *Rumors: Uses, Interpretation and Necessity*. Routledge, 2017. DOI: 10.4324/9781315128801 3

[64] Hamid Karimi and Jiliang Tang. Learning hierarchical discourse-level structure for fake news detection. *ArXiv Preprint ArXiv:1903.07389*, 2019. 11, 12

[65] Jooyeon Kim, Behzad Tabibian, Alice Oh, Bernhard Schölkopf, and Manuel Gomez-Rodriguez. Leveraging the crowd to detect and reduce the spread of fake news and misinformation. In *Proc. of the 11th ACM International Conference on Web Search and Data Mining*, pages 324–332, 2018. DOI: 10.1145/3159652.3159734 30

[66] Benjamin King. Step-wise clustering procedures. *Journal of the American Statistical Association*, 62(317):86–101, 1967. DOI: 10.2307/2282912 16

[67] Diederik P. Kingma and Max Welling. Auto-encoding variational Bayes. *ArXiv Preprint ArXiv:1312.6114*, 2013. 86

[68] Thomas N. Kipf and Max Welling. Semi-supervised classification with graph convolutional networks. In *ICLR*, 2017. 48

[69] Angelika Kirilin and Micheal Strube. Exploiting a speaker's credibility to detect fake news. In *Proc. of Data Science, Journalism and Media Workshop at KDD, (DSJM)*, 2018. 12

[70] Günter Klambauer, Thomas Unterthiner, Andreas Mayr, and Sepp Hochreiter. Self-normalizing neural networks. In *Advances in Neural Information Processing Systems*, pages 971–980, 2017. 49

[71] David O. Klein and Joshua R. Wueller. Fake news: A legal perspective. 2017. https://papers.ssrn.com/sol3/papers.cfm?abstract_id=2958790 2

[72] Hanna Köpcke and Erhard Rahm. Frameworks for entity matching: A comparison. *Data and Knowledge Engineering*, 2010. DOI: 10.1016/j.datak.2009.10.003 23

[73] Danai Koutra, Tai-You Ke, U. Kang, Duen Horng Polo Chau, Hsing-Kuo Kenneth Pao, and Christos Faloutsos. Unifying guilt-by-association approaches: Theorems and fast algorithms. In *Joint European Conference on Machine Learning and Knowledge Discovery in Databases*, pages 245–260, Springer, 2011. DOI: 10.1007/978-3-642-23783-6_16 63

[74] Juhi Kulshrestha, Motahhare Eslami, Johnnatan Messias, Muhammad Bilal Zafar, Saptarshi Ghosh, Krishna P. Gummadi, and Karrie Karahalios. Quantifying search bias: Investigating sources of bias for political searches in social media. In *CSCW*, 2017. DOI: 10.1145/2998181.2998321 27

[75] David M. J. Lazer, Matthew A. Baum, Yochai Benkler, Adam J. Berinsky, Kelly M. Greenhill, Filippo Menczer, Miriam J. Metzger, Brendan Nyhan, Gordon Pennycook, David Rothschild, et al. The science of fake news. *Science*, 359(6380):1094–1096, 2018. DOI: 10.1126/science.aao2998 2

[76] Quoc Le and Tomas Mikolov. Distributed representations of sentences and documents. In *International Conference on Machine Learning*, pages 1188–1196, 2014. 44

[77] Tony Lesce. Scan: Deception detection by scientific content analysis. *Law and Order*, 38(8):3–6, 1990. 17

[78] Jiwei Li, Minh-Thang Luong, and Dan Jurafsky. A hierarchical neural autoencoder for paragraphs and documents. *ArXiv Preprint ArXiv:1506.01057*, 2015. DOI: 10.3115/v1/p15-1107 12

[79] Yaliang Li, Jing Gao, Chuishi Meng, Qi Li, Lu Su, Bo Zhao, Wei Fan, and Jiawei Han. A survey on truth discovery. *ACM SIGKDD Explorations Newsletter*, 17(2):1–16, 2016. DOI: 10.1145/2897350.2897352 23

[80] Yang Liu and Yi-Fang Brook Wu. Early detection of fake news on social media through propagation path classification with recurrent and convolutional networks. In *32nd AAAI Conference on Artificial Intelligence*, 2018. 50, 60, 61

[81] Yunfei Long, Qin Lu, Rong Xiang, Minglei Li, and Chu-Ren Huang. Fake news detection through multi-perspective speaker profiles. In *Proc. of the 8th International Joint Conference on Natural Language Processing (Volume 2: Short Papers)*, vol. 2, pages 252–256, 2017. 12

[82] David G. Lowe. Distinctive image features from scale-invariant key-points. *International Journal of Computer Vision*, 60(2):91–110, 2004. DOI: 10.1023/b:visi.0000029664.99615.94 15

[83] Jiasen Lu, Jianwei Yang, Dhruv Batra, and Devi Parikh. Hierarchical question-image co-attention for visual question answering. In *NIPS*, 2016. 76, 77

[84] Amr Magdy and Nayer Wanas. Web-based statistical fact checking of textual documents. In *Proc. of the 2nd International Workshop on Search and Mining User-Generated Contents*, pages 103–110, ACM, 2010. DOI: 10.1145/1871985.1872002 21

[85] Peter V. Marsden and Noah E. Friedkin. Network studies of social influence. *Sociological Methods and Research*, 22(1):127–151, 1993. DOI: 10.4135/9781452243528.n1 41

[86] Robert R. McCrae, Paul T. Costa, Margarida Pedroso de Lima, António Simões, Fritz Ostendorf, Alois Angleitner, Iris Marušić, Denis Bratko, Gian Vittorio Caprara, Claudio Barbaranelli, et al. Age differences in personality across the adult life span: Parallels in five cultures. *Developmental Psychology*, 35(2):466, 1999. DOI: 10.1037/0012-1649.35.2.466 26

[87] Miller McPherson, Lynn Smith-Lovin, and James M. Cook. Birds of a feather: Homophily in social networks. *Annual Review of Sociology*, 27(1):415–444, 2001. DOI: 10.1146/annurev.soc.27.1.415 41

[88] Nicco Mele, David Lazer, Matthew Baum, Nir Grinberg, Lisa Friedland, Kenneth Joseph, Will Hobbs, and Carolina Mattsson. Combating fake news: An agenda for research and action, 2017. https://shorensteincenter.org/wp-content/uploads/2017/05/Combating-Fake-News-Agenda-for-Research-1.pdf 95

[89] Hugo Mercier. How gullible are we? A review of the evidence from psychology and social science. *Review of General Psychology*, 21(2):103–122, 2017. DOI: 10.1037/gpr0000111 1

[90] Tomas Mikolov, Ilya Sutskever, Kai Chen, Greg S. Corrado, and Jeff Dean. Distributed representations of words and phrases and their compositionality. In *NIPS*, 2013. 11, 37, 56

[91] Tanushree Mitra and Eric Gilbert. Credbank: A large-scale social media corpus with associated credibility annotations. In *ICWSM*, 2015. 80

[92] Federico Monti, Fabrizio Frasca, Davide Eynard, Damon Mannion, and Michael M. Bronstein. Fake news detection on social media using geometric deep learning. *ArXiv Preprint ArXiv:1902.06673*, 2019. 48

[93] Subhabrata Mukherjee and Gerhard Weikum. Leveraging joint interactions for credibility analysis in news communities. In *CIKM*, 2015. DOI: 10.1145/2806416.2806537 23

[94] Danny Murphy. Fake news 101. *Independently Published*, 2019. 95

[95] Eni Mustafaraj and Panagiotis Takis Metaxas. The fake news spreading plague: Was it preventable? *ArXiv Preprint ArXiv:1703.06988*, 2017. DOI: 10.1145/3091478.3091523 2

[96] Mark E. J. Newman. Finding community structure in networks using the eigenvectors of matrices. *Physical Review E*, 74(3):036104, 2006. DOI: 10.1103/physreve.74.036104 44

[97] Maximilian Nickel, Kevin Murphy, Volker Tresp, and Evgeniy Gabrilovich. A review of relational machine learning for knowledge graphs. *Proc. of the IEEE*, 104(1):11–33, 2016. DOI: 10.1109/jproc.2015.2483592 23

[98] Maximilian Nickel, Volker Tresp, and Hans-Peter Kriegel. Factorizing yago: Scalable machine learning for linked data. In *Proc. of the 21st International Conference on World Wide Web*, pages 271–280, ACM, 2012. DOI: 10.1145/2187836.2187874 23

[99] Raymond S. Nickerson. Confirmation bias: A ubiquitous phenomenon in many guises. *Review of General Psychology*, 2(2):175, 1998. DOI: 10.1037/1089-2680.2.2.175 3, 27

[100] Brendan Nyhan and Jason Reifler. When corrections fail: The persistence of political misperceptions. *Political Behavior*, 32(2):303–330, 2010. DOI: 10.1007/s11109-010-9112-2 3

[101] Aude Oliva and Antonio Torralba. Modeling the shape of the scene: A holistic representation of the spatial envelope. *International Journal of Computer Vision*, 42(3):145–175, 2001. DOI: 10.1023/A:1011139631724 15

[102] Ray Oshikawa, Jing Qian, and William Yang Wang. A survey on natural language processing for fake news detection. *ArXiv Preprint ArXiv:1811.00770*, 2018. 9

[103] Evangelos E. Papalexakis and Nicholas D. Sidiropoulos. Co-clustering as multilinear decomposition with sparse latent factors. In *IEEE International Conference on Acoustics, Speech and Signal Processing, (ICASSP)*, pages 2064–2067, 2011. DOI: 10.1109/icassp.2011.5946731 65

[104] Christopher Paul and Miriam Matthews. The Russian firehose of falsehood propaganda model. http://www.intgovforum.org/multilingual/sites/default/fil es/webform/RAND_PE198.pdf 5

[105] James W. Pennebaker, Ryan L. Boyd, Kayla Jordan, and Kate Blackburn. The development and psychometric properties of LIWC2015. *Technical Report*, 2015. DOI: 10.15781/T29G6Z 11

[106] Jeffrey Pennington, Richard Socher, and Christopher Manning. Glove: Global vectors for word representation. In *Proc. of the Conference on Empirical Methods in Natural Language Processing, (EMNLP)*, pages 1532–1543, 2014. DOI: 10.3115/v1/d14-1162 44

[107] Verónica Pérez-Rosas, Bennett Kleinberg, Alexandra Lefevre, and Rada Mihalcea. Automatic detection of fake news. *ArXiv Preprint ArXiv:1708.07104*, 2017. 9

[108] Bryan Perozzi, Rami Al-Rfou, and Steven Skiena. Deepwalk: Online learning of social representations. In *Proc. of the 20th ACM SIGKDD International Conference on Knowledge Discovery and Data Mining*, pages 701–710, 2014. DOI: 10.1145/2623330.2623732 43

[109] Trung Tin Pham. A study on deep learning for fake news detection. 2018. `https://150.65.5.203/dspace/bitstream/10119/15196/3/paper.pdf` 12

[110] Kashyap Popat, Subhabrata Mukherjee, Andrew Yates, and Gerhard Weikum. Declare: Debunking fake news and false claims using evidence-aware deep learning. *ArXiv Preprint ArXiv:1809.06416*, 2018. 69, 70

[111] Martin Potthast, Johannes Kiesel, Kevin Reinartz, Janek Bevendorff, and Benno Stein. A stylometric inquiry into hyperpartisan and fake news. *ArXiv Preprint ArXiv:1702.05638*, 2017. 2, 9

[112] Feng Qian, Chengyue Gong, Karishma Sharma, and Yan Liu. Neural user response generator: Fake news detection with collective user intelligence. *Proc. of the 27th International Joint Conference on Artificial Intelligence*, pages 3834–3840, AAAI Press, 2018. DOI: 10.24963/ijcai.2018/533 56, 79

[113] Walter Quattrociocchi, Antonio Scala, and Cass R. Sunstein. Echo chambers on Facebook. *SSRN 2795110*, 2016. DOI: 10.2139/ssrn.2795110 5

[114] Hannah Rashkin, Eunsol Choi, Jin Yea Jang, Svitlana Volkova, and Yejin Choi. Truth of varying shades: Analyzing language in fake news and political fact-checking. In *Proc. of the Conference on Empirical Methods in Natural Language Processing*, pages 2931–2937, 2017. DOI: 10.18653/v1/d17-1317 11

[115] Benjamin Riedel, Isabelle Augenstein, Georgios P. Spithourakis, and Sebastian Riedel. A simple but tough-to-beat baseline for the fake news challenge stance detection task. *ArXiv Preprint ArXiv:1707.03264*, 2017. 95

[116] Victoria L. Rubin, Yimin Chen, and Niall J. Conroy. Deception detection for news: Three types of fakes. *Proc. of the Association for Information Science and Technology*, 52(1):1–4, 2015. DOI: 10.1002/pra2.2015.145052010083 2

[117] Victoria L. Rubin, Niall J. Conroy, Yimin Chen, and Sarah Cornwell. Fake news or truth? Using satirical cues to detect potentially misleading news. In *Proc. of NAACL-HLT*, pages 7–17, 2016. DOI: 10.18653/v1/w16-0802 2

[118] Victoria L. Rubin and Tatiana Lukoianova. Truth and deception at the rhetorical structure level. *Journal of the Association for Information Science and Technology*, 66(5):905–917, 2015. DOI: 10.1002/asi.23216 17

[119] Natali Ruchansky, Sungyong Seo, and Yan Liu. CSI: A hybrid deep model for fake news. *ArXiv Preprint ArXiv:1703.06959*, 2017. 11, 44, 45, 51

[120] Giovanni C. Santia and Jake Ryland Williams. Buzzface: A news veracity dataset with Facebook user commentary and egos. In *ICWSM*, 2018. 80

[121] Maarten Sap, Gregory Park, Johannes Eichstaedt, Margaret Kern, David Stillwell, Michal Kosinski, Lyle Ungar, and Hansen Andrew Schwartz. Developing age and gender predictive lexica over social media. In *EMNLP*, 2014. DOI: 10.3115/v1/d14-1121 26

[122] H. Andrew Schwartz, Johannes C. Eichstaedt, Margaret L. Kern, Lukasz Dziurzynski, Stephanie M. Ramones, Megha Agrawal, Achal Shah, Michal Kosinski, David Stillwell, Martin E. P. Seligman, et al. Personality, gender, and age in the language of social media: The open-vocabulary approach. *PloS One*, 8(9):e73791, 2013. DOI: 10.1371/journal.pone.0073791 26

[123] Norbert Schwarz, Eryn Newman, and William Leach. Making the truth stick and the myths fade: Lessons from cognitive psychology. *Behavioral Science and Policy*, 2(1):85–95, 2016. DOI: 10.1353/bsp.2016.0009 3

[124] Jingbo Shang, Jiaming Shen, Tianhang Sun, Xingbang Liu, Anja Gruenheid, Flip Korn, Ádám D. Lelkes, Cong Yu, and Jiawei Han. Investigating rumor news using agreement-aware search. In *Proc. of the 27th ACM International Conference on Information and Knowledge Management*, pages 2117–2125, 2018. DOI: 10.1145/3269206.3272020 95

[125] Chengcheng Shao, Giovanni Luca Ciampaglia, Alessandro Flammini, and Filippo Menczer. Hoaxy: A platform for tracking online misinformation. In *WWW*, 2016. DOI: 10.1145/2872518.2890098 85, 86, 90

[126] Chengcheng Shao, Giovanni Luca Ciampaglia, Onur Varol, Kai-Cheng Yang, Alessandro Flammini, and Filippo Menczer. The spread of low-credibility content by social bots. *Nature Communications*, 9(1):4787, 2018. DOI: 10.1038/s41467-018-06930-7 6

[127] Chengcheng Shao, Giovanni Luca Ciampaglia, Onur Varol, Kaicheng Yang, Alessandro Flammini, and Filippo Menczer. The spread of low-credibility content by social bots. *ArXiv Preprint ArXiv:1707.07592*, 2017. DOI: 10.1038/s41467-018-06930-7 49

[128] Karishma Sharma, Feng Qian, He Jiang, Natali Ruchansky, Ming Zhang, and Yan Liu. Combating fake news: A survey on identification and mitigation techniques. *ArXiv Preprint ArXiv:1901.06437*, 2019. DOI: 10.1145/3305260 79

[129] Baoxu Shi and Tim Weninger. Fact checking in heterogeneous information networks. In *WWW*, 2016. DOI: 10.1145/2872518.2889354 22, 23

[130] Prashant Shiralkar, Alessandro Flammini, Filippo Menczer, and Giovanni Luca Ciampaglia. Finding streams in knowledge graphs to support fact checking. *ArXiv Preprint ArXiv:1708.07239*, 2017. DOI: 10.1109/icdm.2017.105 22

[131] Kai Shu, H. Russell Bernard, and Huan Liu. Studying fake news via network analysis: Detection and mitigation. *CoRR*, abs/1804.10233, 2018. DOI: 10.1007/978-3-319-94105-9_3 79

[132] Kai Shu, Limeng Cui, Suhang Wang, Dongwon Lee, and Huan Liu. Defend: Explainable fake news detection. In *Proc. of the 25th ACM SIGKDD International Conference on Knowledge Discovery and Data Mining*, 2019. 71, 73

[133] Kai Shu, Deepak Mahudeswaran, and Huan Liu. Fakenewstracker: A tool for fake news collection, detection, and visualization. *Computational and Mathematical Organization Theory*, pages 1–12, 2018. DOI: 10.1007/s10588-018-09280-3 51, 86, 88

[134] Kai Shu, Deepak Mahudeswaran, Suhang Wang, Dongwon Lee, and Huan Liu. Fakenewsnet: A data repository with news content, social context and dynamic information for studying fake news on social media. *ArXiv Preprint ArXiv:1809.01286*, 2018. 26, 79

[135] Kai Shu, Deepak Mahudeswaran, Suhang Wang, and Huan Liu. Hierarchical propagation networks for fake news detection: Investigation and exploitation. *ArXiv Preprint ArXiv:1903.09196*, 2019. 49, 53

[136] Kai Shu, Amy Sliva, Suhang Wang, Jiliang Tang, and Huan Liu. Fake news detection on social media: A data mining perspective. *ACM SIGKDD Explorations Newsletter*, 19(1):22–36, 2017. DOI: 10.1145/3137597.3137600 2, 43, 44, 72

[137] Kai Shu, Suhang Wang, Thai Le, Dongwon Lee, and Huan Liu. Deep headline generation for clickbait detection. In *IEEE International Conference on Data Mining, (ICDM)*, pages 467–476, 2018. DOI: 10.1109/icdm.2018.00062 17

[138] Kai Shu, Suhang Wang, and Huan Liu. Exploiting tri-relationship for fake news detection. *ArXiv Preprint ArXiv:1712.07709*, 2017. 10

[139] Kai Shu, Suhang Wang, and Huan Liu. Understanding user profiles on social media for fake news detection. In *IEEE Conference on Multimedia Information Processing and Retrieval, (MIPR)*, pages 430–435, 2018. DOI: 10.1109/mipr.2018.00092 26

[140] Kai Shu, Suhang Wang, and Huan Liu. Understanding user profiles on social media for fake news detection. In *1st IEEE International Workshop on "Fake MultiMedia," (FakeMM)*, 2018. DOI: 10.1109/mipr.2018.00092 46

[141] Kai Shu, Suhang Wang, and Huan Liu. Beyond news contents: The role of social context for fake news detection. In *Proc. of the 12th ACM International Conference on Web Search and Data Mining*, pages 312–320, 2019. DOI: 10.1145/3289600.3290994 41, 47

[142] Kai Shu, Xinyi Zhou, Suhang Wang, Reza Zafarani, and Huan Liu. The role of user profiles for fake news detection. *ArXiv Preprint ArXiv:2671079*, 2013. 25, 27

[143] Karen Simonyan and Andrew Zisserman. Very deep convolutional networks for large-scale image recognition. *ArXiv Preprint ArXiv:1409.1556*, 2014. 16

[144] Kihyuk Sohn, Honglak Lee, and Xinchen Yan. Learning structured output representation using deep conditional generative models. In *Advances in Neural Information Processing Systems*, pages 3483–3491, 2015. 57

[145] Nitish Srivastava, Geoffrey Hinton, Alex Krizhevsky, Ilya Sutskever, and Ruslan Salakhutdinov. Dropout: A simple way to prevent neural networks from overfitting. *The Journal of Machine Learning Research*, 15(1):1929–1958, 2014. 16

[146] Stefan Stieglitz and Linh Dang-Xuan. Emotions and information diffusion in social media—sentiment of microblogs and sharing behavior. *Journal of Management Information Systems*, 29(4):217–248, 2013. DOI: 10.2753/mis0742-1222290408 35

[147] Eugenio Tacchini, Gabriele Ballarin, Marco L. Della Vedova, Stefano Moret, and Luca de Alfaro. Some like it hoax: Automated fake news detection in social networks. *ArXiv Preprint ArXiv:1704.07506*, 2017. 34, 80

[148] Henri Tajfel and John C. Turner. An integrative theory of intergroup conflict. *The Social Psychology of Intergroup Relations*, 33(47):74, 1979. 3

[149] Henri Tajfel and John C. Turner. The social identity theory of intergroup behavior. 2004. DOI: 10.4324/9780203505984-16 3

[150] Jian Tang, Meng Qu, Mingzhe Wang, Ming Zhang, Jun Yan, and Qiaozhu Mei. Line: Large-scale information network embedding. In *Proc. of the 24th International Conference on World Wide Web*, pages 1067–1077, International World Wide Web Conferences Steering Committee, 2015. DOI: 10.1145/2736277.2741093 43

[151] Andreas Thor and Erhard Rahm. Moma-a mapping-based object matching system. In *CIDR*, 2007. 23

[152] Sebastian Tschiatschek, Adish Singla, Manuel Gomez Rodriguez, Arpit Merchant, and Andreas Krause. Fake news detection in social networks via crowd signals. In *WWW*, 2018. DOI: 10.1145/3184558.3188722 31, 32

[153] Amos Tversky and Daniel Kahneman. The framing of decisions and the psychology of choice. *Science*, 211(4481):453–458, 1981. DOI: 10.1007/978-1-4613-2391-4_2 3

[154] Amos Tversky and Daniel Kahneman. Advances in prospect theory: Cumulative representation of uncertainty. *Journal of Risk and Uncertainty*, 5(4):297–323, 1992. DOI: 10.1017/cbo9780511803475.004 3

[155] Madeleine Udell, Corinne Horn, Reza Zadeh, Stephen Boyd, et al. Generalized low rank models. *Foundations and Trends® in Machine Learning*, 9(1):1–118, 2016. DOI: 10.1561/2200000055 10

[156] Udo Undeutsch. Beurteilung der glaubhaftigkeit von aussagen. *Handbuch der Psychologie*, 11:26–181, 1967. 17

[157] Emily Van Duyn and Jessica Collier. Priming and fake news: The effects of elite discourse on evaluations of news media. *Mass Communication and Society*, 22(1):29–48, 2019. DOI: 10.1080/15205436.2018.1511807 3

[158] Andreas Vlachos and Sebastian Riedel. Fact checking: Task definition and dataset construction. *ACL*, 2014. DOI: 10.3115/v1/w14-2508 20

[159] Soroush Vosoughi, Deb Roy, and Sinan Aral. The spread of true and false news online. *Science*, 359(6380):1146–1151, 2018. DOI: 10.1126/science.aap9559 49

[160] Aldert Vrij. Criteria-based content analysis: A qualitative review of the first 37 studies. *Psychology, Public Policy, and Law*, 11(1):3, 2005. DOI: 10.1037/1076-8971.11.1.3 17

[161] Meng Wang, Kuiyuan Yang, Xian-Sheng Hua, and Hong-Jiang Zhang. Towards a relevant and diverse search of social images. *IEEE Transactions on Multimedia*, 12(8):829–842, 2010. DOI: 10.1109/tmm.2010.2055045 15

[162] Suhang Wang and Huan Liu. Deep learning for feature representation. *Feature Engineering for Machine Learning and Data Analytics*, page 279, 2018. DOI: 10.1201/9781315181080-11 16

[163] William Yang Wang. "Liar, liar pants on fire": A new benchmark dataset for fake news detection. *ArXiv Preprint ArXiv:1705.00648*, 2017. DOI: 10.18653/v1/p17-2067 11, 12, 13, 79

[164] Xiao Wang, Peng Cui, Jing Wang, Jian Pei, Wenwu Zhu, and Shiqiang Yang. Community preserving network embedding. In *AAAI*, pages 203–209, 2017. 43

[165] Yaqing Wang, Fenglong Ma, Zhiwei Jin, Ye Yuan, Guangxu Xun, Kishlay Jha, Lu Su, and Jing Gao. EANN: Event adversarial neural networks for multi-modal fake news detection. In *Proc. of the 24th ACM SIGKDD International Conference on Knowledge Discovery and Data Mining*, pages 849–857, 2018. DOI: 10.1145/3219819.3219903 16, 57, 58

[166] Andrew Ward, L. Ross, E. Reed, E. Turiel, and T. Brown. Naive realism in everyday life: Implications for social conflict and misunderstanding. *Values and Knowledge*, pages 103–135, 1997. 3

[167] Gerhard Weikum. What computers should know, shouldn't know, and shouldn't believe. In *WWW*, 2017. DOI: 10.1145/3041021.3051120 23

[168] Janyce Wiebe, Theresa Wilson, and Claire Cardie. Annotating expressions of opinions and emotions in language. *Language Resources and Evaluation*, 39(2–3):165–210, 2005. DOI: 10.1007/s10579-005-7880-9 36

[169] Liang Wu and Huan Liu. Tracing fake-news footprints: Characterizing social media messages by how they propagate. *Proc. of the 11th ACM International Conference on Web Search and Data Mining*, pages 637–645, 2018. DOI: 10.1145/3159652.3159677 44, 50

[170] Liang Wu, Fred Morstatter, and Huan Liu. SlangSD: Building and using a sentiment dictionary of slang words for short-text sentiment classification. *ArXiv Preprint ArXiv:1608.05129*, 2016. 36

[171] You Wu, Pankaj K. Agarwal, Chengkai Li, Jun Yang, and Cong Yu. Toward computational fact-checking. *VLDB*, 7(7):589–600, 2014. DOI: 10.14778/2732286.2732295 22

[172] Caiming Xiong, Victor Zhong, and Richard Socher. Dynamic coattention networks for question answering. *ArXiv Preprint ArXiv:1611.01604*, 2016. 76

[173] Wei Xu, Xin Liu, and Yihong Gong. Document clustering based on non-negative matrix factorization. In *Proc. of the 26th Annual International ACM SIGIR Conference on Research and Development in Informaion Retrieval*, pages 267–273, 2003. DOI: 10.1145/860435.860485 10

[174] Shuo Yang, Kai Shu, Suhang Wang, Renjie Gu, Fan Wu, and Huan Liu. Unsupervised fake news detection on social media: A generative approach. *Proc. of 33rd AAAI Conference on Artificial Intelligence*, 2019. 65, 66

[175] Yang Yang, Lei Zheng, Jiawei Zhang, Qingcai Cui, Zhoujun Li, and Philip S. Yu. Ti-CNN: Convolutional neural networks for fake news detection. *ArXiv Preprint ArXiv:1806.00749*, 2018. 11

[176] Yuting Yang, Juan Cao, Mingyan Lu, Jintao Li, and Chia-Wen Lin. How to write high-quality news on social network? Predicting news quality by mining writing style. *ArXiv Preprint ArXiv:1902.04231*, 2019. 19

[177] Zichao Yang, Diyi Yang, Chris Dyer, Xiaodong He, Alex Smola, and Eduard Hovy. Hierarchical attention networks for document classification. In *Proc. of the Conference of the North American Chapter of the Association for Computational Linguistics: Human Language Technologies*, pages 1480–1489, 2016. DOI: 10.18653/v1/n16-1174 12, 74

[178] Tom Young, Devamanyu Hazarika, Soujanya Poria, and Erik Cambria. Recent trends in deep learning based natural language processing. *IEEE Computational Intelligence Magazine*, 13(3):55–75, 2018. DOI: 10.1109/mci.2018.2840738 11

[179] Reza Zafarani, Mohammad Ali Abbasi, and Huan Liu. *Social Media Mining: An Introduction*. Cambridge University Press, 2014. DOI: 10.1017/cbo9781139088510 1

[180] Robert B. Zajonc. Attitudinal effects of mere exposure. *Journal of Personality and Social Psychology*, 9(2p2):1, 1968. DOI: 10.1037/h0025848 5

[181] Robert B. Zajonc. Mere exposure: A gateway to the subliminal. *Current Directions in Psychological Science*, 10(6):224–228, 2001. DOI: 10.1017/cbo9780511618031.026 5

[182] Qiang Zhang, Aldo Lipani, Shangsong Liang, and Emine Yilmaz. Reply-aided detection of misinformation via Bayesian deep learning. In *Companion Proceedings of the Web Conference*, 2019. DOI: 10.1145/3308558.3313718 12

[183] Xichen Zhang and Ali A. Ghorbani. An overview of online fake news: Characterization, detection, and discussion. *Information Processing and Management*, 2019. DOI: 10.1016/j.ipm.2019.03.004 79

[184] Xing Zhou, Juan Cao, Zhiwei Jin, Fei Xie, Yu Su, Dafeng Chu, Xuehui Cao, and Junqiang Zhang. Real-time news certification system on sina weibo. In *Proc. of the 24th International Conference on World Wide Web*, pages 983–988, ACM, 2015. DOI: 10.1145/2740908.2742571 92, 93

[185] Xinyi Zhou and Reza Zafarani. Fake news: A survey of research, detection methods, and opportunities. *ArXiv Preprint ArXiv:1812.00315*, 2018. 3, 22

[186] Xinyi Zhou, Reza Zafarani, Kai Shu, and Huan Liu. Fake news: Fundamental theories, detection strategies and challenges. In *Proc. of the 12th ACM International Conference on Web Search and Data Mining*, pages 836–837, 2019. DOI: 10.1145/3289600.3291382 2

[187] Xinyi Zhou, Reza Zafarani, Kai Shu, and Huan Liu. Fake news: Fundamental theories, detection strategies and challenges. In *WSDM*, 2019. DOI: 10.1145/3289600.3291382 96

Authors' Biographies

KAI SHU

Kai Shu is a Ph.D. student and research assistant at the Data Mining and Machine Learning (DMML) Lab at Arizona State University. He received his B.S./M.S. from Chongqing University in 2012 and 2015, respectively. His research interests include fake news detection, social computing, data mining, and machine learning. He was awarded ASU CIDSE Doctorial Fellowship 2015, the 1st place of SBP Disinformation Challenge 2018, University Graduate Fellowship, and various scholarships. He co-presented two tutorials in KDD 2019 and WSDM 2019, and has published innovative works in highly ranked journals and top conference proceedings such as ACM KDD, WSDM, WWW, CIKM, IEEE ICDM, IJCAI, AAAI, and MICCAI. He also worked as a research intern at Yahoo! Research and Microsoft Research in 2018 and 2019, respectively. More can be found at http://www.public.asu.edu/~skai2/.

HUAN LIU

Huan Liu is a professor of Computer Science and Engineering at Arizona State University. Before he joined ASU, he worked at Telecom Australia Research Labs and was on the faculty at the National University of Singapore. His research interests are in data mining, machine learning, social computing, and artificial intelligence, investigating interdisciplinary problems that arise in many real-world, data-intensive applications with high-dimensional data of disparate forms such as social media. He is a co-author of *Social Media Mining: An Introduction* (Cambridge University Press). He is a founding organizer of the International Conference Series on Social Computing, Behavioral-Cultural Modeling, and Prediction, Field Chief Editor of *Frontiers in Big Data* and its Specialty Chief Editor in *Data Mining and Management*. He is a Fellow of ACM, AAAI, AAAS, and IEEE. More can be found about him at http://www.public.asu.edu/~huanliu.

Printed in the United States
by Baker & Taylor Publisher Services